Springer Theses

Recognizing Outstanding Ph.D. Research

Aims and Scope

The series "Springer Theses" brings together a selection of the very best Ph.D. theses from around the world and across the physical sciences. Nominated and endorsed by two recognized specialists, each published volume has been selected for its scientific excellence and the high impact of its contents for the pertinent field of research. For greater accessibility to non-specialists, the published versions include an extended introduction, as well as a foreword by the student's supervisor explaining the special relevance of the work for the field. As a whole, the series will provide a valuable resource both for newcomers to the research fields described, and for other scientists seeking detailed background information on special questions. Finally, it provides an accredited documentation of the valuable contributions made by today's younger generation of scientists.

Theses are accepted into the series by invited nomination only and must fulfill all of the following criteria

- They must be written in good English.
- The topic should fall within the confines of Chemistry, Physics, Earth Sciences, Engineering and related interdisciplinary fields such as Materials, Nanoscience, Chemical Engineering, Complex Systems and Biophysics.
- The work reported in the thesis must represent a significant scientific advance.
- If the thesis includes previously published material, permission to reproduce this must be gained from the respective copyright holder.
- They must have been examined and passed during the 12 months prior to nomination.
- Each thesis should include a foreword by the supervisor outlining the significance of its content.
- The theses should have a clearly defined structure including an introduction accessible to scientists not expert in that particular field.

More information about this series at http://www.springer.com/series/8790

Yuan Dong

Dynamical Analysis of Non-Fourier Heat Conduction and Its Application in Nanosystems

Doctoral Thesis accepted by
Tsinghua University, Beijing, China

Springer

Author
Dr. Yuan Dong
Key Laboratory for Thermal Science and Power Engineering of Ministry of Education
Department of Engineering Mechanics
Tsinghua University
Beijing
China

Supervisor
Prof. Zengyuan Guo
Key Laboratory for Thermal Science and Power Engineering of Ministry of Education
Department of Engineering Mechanics
Tsinghua University
Beijing
China

ISSN 2190-5053 ISSN 2190-5061 (electronic)
Springer Theses
ISBN 978-3-662-48483-8 ISBN 978-3-662-48485-2 (eBook)
DOI 10.1007/978-3-662-48485-2

Library of Congress Control Number: 2015950440

Springer Heidelberg New York Dordrecht London

Printed on acid-free paper

Springer-Verlag GmbH Berlin Heidelberg is part of Springer Science+Business Media
(www.springer.com)

Parts of this thesis have been published in the following journal articles:

[1] Y. Dong, B.Y. Cao and Z. Y. Guo. Ballistic–diffusive phonon transport and size induced anisotropy of thermal conductivity of silicon nanofilms. Physica E: Low-dimensional Systems and Nanostructures, 66, 1 (2015).

[2] Y. Dong and Z. Y. Guo. Hydrodynamic modeling of heat conduction in nanoscale systems. Journal of Nanoscience and Nanotechnology, 15, 3229 (2014).

[3] Y. Dong, B.Y. Cao and Z. Y. Guo. Size dependent thermal conductivity of Si nanosystems based on phonon gas dynamics. Physica E: Low-dimensional Systems and Nanostructures, 56, 256 (2014).

[4] Y. C. Hua, Y. Dong and B. Y. Cao. Monte Carlo simulation of phonon ballistic diffusive heat conduction in silicon nano film. Acta Physica Sinica, 62(24), 244401 (2013). (In Chinese)

[5] Y. Dong, B.Y. Cao and Z. Y. Guo. Temperature in nonequilibrium states and non-Fourier heat conduction. Physical Review E, 87, 032150 (2013).

[6] Y. Dong. Clarification of Onsager Reciprocal Relations Based on Thermomass Theory. Physical Review E, 86, 062101 (2012).

[7] Y. Dong, B.Y. Cao and Z. Y. Guo. General expression for entropy production in transport processes based on the thermomass model. Physical Review E, 85, 061107 (2012)

[8] Y. Dong and Z. Y. Guo. The modification of entropy production by heat condution in non-equilibrium thermodynamics. Acta Physica Sinica, 61(3), 030507 (2012). (In Chinese)

[9] Y. Dong and Z. Y. Guo. From Thermomass Theory to Kinetomass Theory. Journal of Engineering Thermophysics, 33, 465 (2012). (In Chinese)

[10] Y. Dong, B. Y. Cao and Z. Y. Guo. Generalized heat conduction laws based on thermomass theory and phonon hydrodynamics. Journal of Applied Physics, 110, 063504 (2011).

[11] Y. Dong and Z. Y. Guo. Entropy analyses for hyperbolic heat conduction based on the thermomass model. International Journal of Heat and Mass Transfer, 54, 1924 (2011).

[12] Y. Dong and Z. Y. Guo. Entropy Analysis in the Heat Wave Propagation. Journal of Engineering Thermophysics, 32, 1889 (2011). (In Chinese)

To my family,
for their unconditional love
and firm support

Supervisor's Foreword

Heat conduction is a traditional subject that can be traced back to 1822, the year Fourier's conduction law was established. Through nearly two centuries the application of the theory of heat conduction has been immensely developed in the fields of energy, electronics, material processing, environment protection, as well as human health. The heat conduction research based on Fourier's conduction law usually focuses on how to transfer heat efficiently for heating or cooling objects. However, in the past three decades, with the development of short pulse laser and fabrication of nanomaterials, the validation of Fourier's law has been challenged. It was pointed out in the mid-twentieth century that Fourier's law implies an infinite heat propagation speed, a physically unacceptable notion. In studies on ultrafast laser heating on materials from the 1980s, it is observed that the temperature response on laser heating exhibits the behavior of lagging, relaxation, or delay, which indicates the failure of Fourier's conduction law. On the other hand, in low-dimensional materials such as carbon nanotube and graphene, as well as nanosized semiconductors, the heat conduction shows a size-dependent behavior. The limited size of materials can either provide ultrahigh heat conductivity, which sheds light on the heat management of large-scale Integrated circuits, or much-suppressed heat conductivity, which can enhance the figure of merit of thermoelectric devices. These applications provide potential solutions to the emergent needs raised by modern engineering. However, the scientific understanding and modeling of heat conduction in these extreme conductions are far from satisfactory.

In the present work by Dr. Dong, non-Fourier heat conduction is investigated through various perspectives. The basic idea originates from the thermomass theory, which was established by our research group since 2005. In the thermomass theory, based on the mass–energy equivalence of Einstein, thermal energy is regarded as a weighable fluid flowing through the porous mediums, which is different from the caloric theory of the eighteenth century. As a result, a general heat conduction law was presented to describe the relationship between the heat flux and the temperature gradient by use of principle of fluid dynamics, which degenerates to Fourier's conduction law or other non-Fourier heat conduction models under different simplifications. Dr. Dong's work first studies the microscopic foundation of thermomass theory in the dielectric medium, where the main heat carriers are phonons. Based on the phonon Boltzmann theory, he revealed the connection between the phonon quasi-momentum and the real momentum of phonon gas. In this way, the momentum balance equation of phonon gas can be formulated, which then leads to the general heat conduction law beyond Fourier's conduction law. The general heat conduction law is similar to the phonon hydrodynamics model proposed in the 1960s, with a new term corresponding to the convection effect of phonon gas. The author proves that this difference comes from the higher order expansion of the phonon distribution function. This derivation bridges the microscopic and macroscopic theories. It not only provides a microscopic explanation for the thermomass theory, but also clarifies the hierarchy for many non-Fourier models.

Second, the thermomass theory enables one to analysis the irreversible thermodynamics from a perspective of fluid mechanics. By distinction of the reversible and irreversible effects in the general heat conduction law, this work claims that irreversibility in non-Fourier heat conduction is induced only by the friction force rather than the driving force. Thus the traditional expression of entropy production has to be modified. Like the analysis of the extended irreversible thermodynamics, the proposed general entropy production avoids the negativity paradox in non-Fourier heat conduction processes. The modification of entropy production naturally causes the revision of the entropy and temperature in thermodynamics. Using the approach of compressible fluid dynamics, the author announces the static temperature and total temperature in non-Fourier heat conduction, which are the static and total pressures of the phonon gas. The distinction between these two temperatures is comprehensively investigated through the thermodynamic laws, as well as the phonon Boltzmann equation. The by-product of the above analysis is that the generalized forces and fluxes in the entropy production should be the real forces and fluxes of the thermomass flow. With this discovery, the long-existing problem in the derivation of Onsager reciprocal relation, namely the generalized fluxes cannot be expressed by the time derivatives of state variables, is solved. The author shows that the time derivative term in Onsager's derivation should be the inertia force of heat conduction. Thus the state variables are formulated as the "displacement of heat," which is the average displacement of transported quantities during fluctuation. The author further provides a macroscopic derivation of the Onsager reciprocal relation based on the principles of Galilean invariance and the third law of Newtonian dynamics.

Lastly, the thermomass theory is used in up-to-date applications, i.e., the nanoscale non-Fourier heat conduction. The size dependence of the effective thermal conductivity in nanosystems is induced by the boundary scattering of heat carriers. In this work, the boundary effect is modeled by the additional boundary friction term raised by the phonon gas viscosity, in analogy to the Brinkman extension for the porous flow. On the other hand, the confined structure also causes the rarefaction effect which reduces the effective viscosity of phonon gas. By accounting for both the viscosity and rarefaction effects the author builds prediction models for the effective thermal conductivity of nanosystems, which agree well with the experiments. Moreover, a ballistic-diffusive model is proposed for the cross-plane thermal conductivity of nanofilms. The author shows that the different heat conduction directions will cause size-dependent heterogeneity of thermal conductivity, which is led by the different geometry confinement mechanisms.

This work manifests the excellent analysis skill, physical insights, and broad knowledge of the author, from the condensed physics to thermodynamics, from fundamental theory to cutting edge applications. It received unanimous high praise from the thesis reviewers. As the supervisor of Dr. Dong, I am glad to recommend this thesis to readers, particularly those specialized or interested in the heat conduction theory, nanotechnology, and thermodynamics.

Beijing, China
August 2015

Prof. Zengyuan Guo

Abstract

Heat conduction cannot be characterized by Fourier's law in extreme conditions such as ultrafast transient heating or nanoscale heat conduction, which is called non-Fourier heat conduction. Based on Einstein's mass–energy equivalence, Guo et al. proposed that thermal energy has its equivalent mass, namely thermomass. Heat conduction is actually the motion of thermomass, which obeys Newton's law of motion. Therefore, the non-Fourier heat conduction can be analyzed from a dynamical viewpoint, which establishes the general heat conduction law with a clear macroscopic physical picture.

Phonons are the main thermal energy carriers in dielectric solids. This work obtains the microscopic foundation of general heat conduction law through the phonon Boltzmann equation. The transient and spatial inertial terms of thermomass come from the first and second orders of expansion of the phonon distribution function, respectively. Neglecting all the high order expansions of the phonon distribution is equivalent to neglecting the inertia terms of thermomass, and reduces the general heat conduction law to Fourier's law. The inertial effect of thermomass cannot be neglected in ultrasmall time or spatial scales, causing non-Fourier heat conduction.

The entropy production in irreversible thermodynamics is the product of generalized forces and fluxes. The classical expression of entropy production is non-positive definite in non-Fourier heat conduction, which violates the second law of thermodynamics. This work defines the real forces and fluxes in heat conduction based on the thermomass theory, instead of the phenomenological generalized forces and fluxes. The forces in entropy production should be the friction forces rather than the driving forces. Therefore, the general expression of entropy production is obtained and is compatible with the non-Fourier heat conduction.

The definition of temperature needs to be modified in non-Fourier heat conduction. This work derives the expressions of static and stagnant temperatures based on the Bernoulli equation of thermomass flow. The static temperature is the true state variable and is consistent with the nonequilibrium temperature in extended irreversible thermodynamics. The internal energy and entropy should be expressed with the static temperature.

The linear regression of fluctuation is assumed in the proof of the Onsager reciprocal relation. It requires that the generalized fluxes are the time derivative of state variables, which is hardly satisfied by usually defined fluxes. This work proves that the linear regression of fluctuation is the balance of inertia and friction forces of the thermomass, which is a non-Fourier heat conduction process. The corresponding state variable of heat flux is the average displacement of thermomass during fluctuation.

The boundaries impose additional resistances on heat conduction in nanosystems, causing the size effect of the effective thermal conductivity. This work adds a viscous term of thermomass in the general heat conduction law to describe the boundary resistance. The in-plane effective thermal conductivities of nanofilms and nanowires are predicted by considering both the nonuniform heat flux profile in the cross section due to phonon gas viscosity and the rarefaction of phonon gas. For cross-plane effective thermal conductivity of nanofilms, a ballistic-diffusive model is built based on the Boltzmann equation regarding the nonequilibrium distribution in near boundary region. These models agree well with the experimental and numerical simulation results.

Acknowledgments

I would like to express my sincere gratitude and appreciation to my advisor, Prof. Zengyuan Guo. His unremitting pursuit of truth sets me an example of a true scholar. His passion, guidance, and inspiration will be the lifelong precious wealth for me.

I am also indebted to Profs. Bingyang Cao, Xing Zhang, Xingang Liang, Moran Wang, and Qunchen for their invaluable help during my Ph.D. career. The discussions with my labmates, including Xiaowei Li, Jing Wu, Quanwen Hou, Xuetao Cheng, Haidong Wang, Weigang Ma, Ruoyu Dong, and Guojie Hu, were very beneficial for my work.

The guidance and advice from Prof. David Jou at Universitat Autònoma de Barcelona and Prof. Gang Chen at MIT during my research visits are much appreciated.

I would like to acknowledge my family members, my father Aiguo Dong and mother Yu Tang, and my beloved wife Xiaoyu Zhou. Their long-lasting love and support across so many years are indispensable to me and all my achievements.

The work of this thesis is financially supported by the National Natural Science Foundation of China (Nos. 51136001, 51356001), Tsinghua University Initiative Scientific Research Program and Tsinghua Scholarship for Overseas Graduate Studies.

Contents

Chapter 1
Introduction

Abstract The traditional Fourier's law of heat conduction is not applicable in ultrafast and ultrasmall conditions. Non-Fourier models thus have been developed to predict these anomalous heat conductions. This chapter reviews the present non-Fourier heat conduction theories. For ultrafast heat conduction, modification models such as the Cattaneo–Vernotte model, dual phase lag model, and hyperbolic two-step model are developed. The common feature of these models is adding the relaxation terms in the traditional Fourier's law. For the steady non-Fourier heat conduction in nanosystems, the size effect of the thermal conductivity has been modeled from the phonon-boundary scattering perspective. On the other hand, the combination of non-Fourier conduction models with irreversible thermodynamics will give the negative entropy production, which violates the second law. The extended irreversible thermodynamics modifies the entropy production by extending the category of state variables to mend this paradox. At the end of this chapter, the approach and main aim of this work are presented.

1.1 Transient Non-Fourier Heat Conduction

In 1822 Fourier [1] proposed the well-known heat conduction law, namely that the heat flux passing through a material is proportional to the local gradient of temperature

$$\boldsymbol{q} = -\kappa \nabla T \tag{1.1}$$

The parameter, κ, is the thermal conductivity. For heat conduction processes without internal heat source or sink, the conservation of internal energy gives

$$\frac{\partial \rho C_{\mathrm{V}} T}{\partial t} + \nabla \cdot \boldsymbol{q} = 0, \tag{1.2}$$

Y. Dong, *Dynamical Analysis of Non-Fourier Heat Conduction and Its Application in Nanosystems*, Springer Theses,
DOI 10.1007/978-3-662-48485-2_1

where ρ is the density of material, C_V is the volumetric specific heat, t is the time. A combination of Eqs. 1.1 and 1.2 yields the temperature evolution relation

$$\frac{\partial T}{\partial t} = \alpha \nabla^2 T, \tag{1.3}$$

where $\alpha = \kappa/\rho C_V$ is the thermal diffusivity. Equation 1.3 indicates a parabolic evolution of temperature. It implies that a sudden temperature disturbance is felt simultaneously by the whole part of medium, in other words, the unphysical infinite propagation speed of thermal disturbance. Nevertheless, Fourier's law can be regarded as a good approximation of the real physics in most traditional engineering cases.

Researchers began to pay attention to this paradox from around 1950. Cattaneo, Morse, and Vernotte [2–4] elucidated that for heat conduction in gases, under the imposed sudden temperature disturbance, a certain amount of time is needed to accelerate the carriers of thermal energy. Therefore, a time delay should exist between the heat flux and the temperature gradient. They proposed the modification of the Fourier's law, i.e., CV model

$$\tau \frac{\partial}{\partial t} \boldsymbol{q} + \boldsymbol{q} = -\kappa \nabla T, \tag{1.4}$$

where τ is the relaxation time. With Eq. 1.2, the CV model gives the hyperbolic temperature evolution equation

$$\tau \frac{\partial^2 T}{\partial t^2} + \frac{\partial T}{\partial t} = \alpha \nabla^2 T, \tag{1.5}$$

which predicts that the heat propagates as an attenuating wave with a speed of $(\alpha/\tau)^{0.5}$.

Although the CV model avoids the paradox of infinite propagation speed, it still has many limitations. For example, the overlap of heat waves predicted by the CV model could lead to the local temperature lower than 0 K [5–7], which is unphysical. On the other hand, many experiments [8–13] indicate that the temperature profile strongly deviates from the CV model in the case of considerable thermal disturbance. Therefore, researchers developed a series of transient non-Fourier heat conduction models beyond the CV models and made comparison with experiments.

In analogy to the relaxational relation between strain and stress in viscoelastic materials, Joseph and Preziosi [14] proposed the Jefferey's type model for heat conduction

$$\tau \frac{\partial}{\partial t} \boldsymbol{q} + \boldsymbol{q} = -\kappa \nabla T - \kappa_1 \frac{\partial}{\partial t} (\nabla T) \tag{1.6}$$

The second term on the right-hand side of Eq. 1.6 suggests that additional mode of heat waves could exist and propagate with a different speed from the primary heat wave. κ_1 is defined as the thermal conductivity corresponding to the "fast mode" of heat wave. The heat waves predicted by the CV model have steep wave fronts, i.e., the discontinuity of temperature profile across the wave front. The heat waves predicted by Eq. 1.6 would have smoother wave fronts with the thickness proportional to $(\kappa_1 x/\kappa_1)^{0.5}$, with x the location of wave front. This behavior is similar to the thickness of shock wave induced by viscosity in gas dynamics.

Guyer and Krumhansl [15, 16] proposed the phonon hydrodynamic model based on the linear solution of phonon Boltzmann equation (BTE)

$$\tau_R \frac{\partial \boldsymbol{q}}{\partial t} + \boldsymbol{q} = -\kappa \nabla T + \frac{\tau_R \tau_N v_s^2}{5} \left(\nabla^2 \boldsymbol{q} + 2\nabla\nabla \cdot \boldsymbol{q} \right), \tag{1.7}$$

where τ_R and τ_N are the characteristic relaxation time of the resistive phonon scattering (R-process) and the normal phonon scattering (N-process), respectively. v_s is the average group velocity of phonons. Due to the second term on its right-hand side, Eq. 1.7 is also able to smooth the heat wave front in a similar manner to Eq. 1.6.

Since the 1980s, the time-domain thermoreflectance (TDTR) based on the femtosecond (fs) laser is widely used to measure the thermophysical properties. Eesley, Fujimoto, and Brorson et al. [17–25] studied the electron–phonon interaction during ultrafast heating, as well as the individual contribution of electron and phonon to the thermal transport in thin metal film. Due to the ultrashort interaction time between laser and metal, the thermal transport cannot be described by the Fourier's law. Qiu and Tien [26–28] analyzed the energy transport in multilayer metal films heated by the fs laser. They proposed that the thermal transport in electron system does not obey the Fourier's law. Instead, the response delays of phonon and electron to laser heating pulses should be considered separately, which leads to the hyperbolic two-step model for ultrafast laser heating on metals.

Özişik and Tzou [29–31] claimed that the interaction of microscopic particles causes macroscopic response delay. If the disturbance is temperature gradient, then the heat flux response will be delayed. If the disturbance is heat flux, then the temperature evolution will be delayed. In this way the relaxation of both heat flux and temperature gradient should be taken into account, which leads to the dual phase lagging (DPL) model

$$q\left(t + \tau_q\right) = -\kappa \nabla T(t + \tau_T), \tag{1.8}$$

where τ_q and τ_T are the characteristic relaxation times of heat flux and temperature gradient. The DPL model can evolve to several relaxational heat conduction models with τ_q and τ_T given specific physical expressions. For example, the DPL model reduces to CV model when $\tau_T = 0$. It converts to the hyperbolic or parabolic two-step models for ultrafast laser–metal interaction when τ_q and τ_T are expressed by the phonon–electron coupling functions. If $\tau_q = \tau_R$ and $\tau_T = 9\tau_N/5$, the DPL model turns to the phonon hydrodynamic model, Eq. 1.7. There are also many works discussing the solutions of DPL model and its microscopic interpretation [32–36].

Chen [37] established the Ballistic-Diffusive model to describe the transient heat conduction based on the BTE. The heat carriers' (e.g., phonons) distribution function in an internal part of the material can be separated into two parts: the ballistic part originating from the boundary and the diffusive part originating from the interior. The ballistic part attenuates exponentially with the distance from the boundary. The diffusive part is approximately governed by the CV model. The ballistic part couples with the diffusive part via the energy conservation equation, which means that the attenuated energy of ballistic carriers is transported to the diffusive ones

$$C_{\mathrm{V}}\left(\tau\frac{\partial^2 T}{\partial t^2}+\frac{\partial T}{\partial t}\right)=\nabla(\kappa\nabla T)-\nabla\cdot\boldsymbol{q}_{\mathrm{b}}, \tag{1.9}$$

where the ballistic heat flux, q_{b}, is determined by the boundary temperature. This model is used to predict the one-dimensional transient heat conduction and agrees well with the numerical solution based on BTE, which prevails over the prediction based on Fourier's law and CV model.

It should be noted that the above non-Fourier heat conduction models are derived in the static solid system. Christov and Jordan [38] indicated that heat wave predicted by the CV model would violate the Galilean invariance in the moving medium. To remove the paradox, they proposed to modify the partial differential operator in the CV model with the objective differential operator D/Dt

$$\frac{\mathrm{D}}{\mathrm{D}t}\boldsymbol{q}=\frac{\partial}{\partial t}q+(\boldsymbol{u}\cdot\nabla)q, \tag{1.10}$$

where u is the moving velocity of heat conduction medium. For the transient heat conduction in gas medium, the flow of gas elements should also be considered. Müller and Ruggeri [39] derived the transient heat conduction model in case of the heat conduction and fluid flow coupling based on the 13-momentum model. In their model the time derivative of heat flux is expressed as

$$\frac{\mathrm{D}}{\mathrm{D}t}\boldsymbol{q}=\frac{\partial}{\partial t}q_{\mathrm{i}}+q_{\mathrm{k}}\frac{\partial u_{\mathrm{i}}}{\partial x_{\mathrm{k}}}-2q_{\mathrm{k}}W_{\mathrm{ik}}, \tag{1.11}$$

where u is the moving velocity of fluid element and W is the angular velocity matrix. Compared with Eqs. 1.10 and 1.11 can be regarded as more a general model which considers not only the translational acceleration but also the Coriolis acceleration.

The experimental measurements of transient non-Fourier conduction in dielectric solids are feasible in low temperature. In such condition the phonon relaxation time is long, which enables the heat wave pulses to stand out of the diffusive processes.

Therefore, one obtains a large signal-to-noise ratio. In 1966, Ackeman et al. [8] reported the second sound (temperature wave) detected in the solid helium at 0.54–0.71 K. They estimated that the propagation speed of heat wave is around 160 m/s. Rogers [9] measured the heat waves in NaF, LiF, and NaI. He claimed the propagation of temperature waves obeys the phonon hydrodynamics model. The effect of τ_N on the heat wave propagation was adopted to estimate the mean free path (MFP) of normal scatterings. The heat waves are also measured in dielectric solids such as Bi and LiF [10–13]. Although the wave-like temperature signals were observed in these works, there are considerable discrepancies between theoretical models and experimental data.

The experimental data for heat waves is inevitably limited by factors such as the purity of samples, contact resistances, and oscillations. The molecular dynamic (MD) simulation can serve as a theoretical tool for investigating transient non-Fourier heat conduction in solids. Tsai and MacDonald [40] used MD simulation to study heat pulses propagation in α-Fe crystals. They indicated that the energy migration in lattices has the feature of wave. The dispersion induces lower propagation speed than the theoretical predictions. The temperature pulses stimulate the translational and longitudinal elastic waves, each accompanied with the second sound wave transmitting at lower speed. Volz et al. [41] studied the transient heat conduction in solid argon with the Lennard-Jones potential. They found that the CV model can accurately predict the heat flux attenuation in the case of small disturbance. With larger disturbance, the prediction by CV model deviates significantly from MD simulation. Thus the CV model with single relaxation time is not suitable to describe the transient heat conduction with larger temperature fluctuation. Shiomi and Maruyama [42] simulated the heat wave propagation in single wall carbon nanotubes (CNTs). The results show that the transient heat transport does not obey Fourier's law. A considerable amount of heat transports as waves. The temperature evolution profile has large differences with the CV model, while is close to the DPL model with two relaxation times. Based on the dispersion curves, they pointed out that the temperature waves are contributed mainly by the optical phonons, since the transport of acoustic phonons is mostly ballistic in this condition.

1.2 Steady Non-Fourier Heat Conduction in Nanosystems

A number of novel nanomaterials are synthesized in recent years, including low-dimensional materials (CNT, graphene) [43–49], nano-semiconductors (silicon nanofilms and nanowires) [50–56], nano-superlattices (GaAs/AlAs, Si/Ge) [57–64], and nano-polymers (Polyethylene nanowires) [65–67]. These nanosized or nanostructured materials exhibit unique chemical, optical, electrical, and thermal properties compared with normal bulk materials and are expected to bring revolutionary breakthroughs in fields such as microelectronics and renewable energy [68–72]. For example, the size of microelectronics is rapidly reducing. Thedensity of large electronic elements causes huge heat flux rate, which becomes the bottleneck in further improvement of integrated circuits (ICs). Low-dimensional materials such as CNT and graphene have much larger heat conductivities compared with traditional materials. They are expected to be made into high performance cooling devices. On the other hand, the significant reduction of thermal conductivities of nano-sized semiconductors raises their thermoelectric figure of merit, which has great potential for harvesting low-temperature waste heat as well as solid-state thermal solar cells.

The non-Fourier heat conduction in nanomaterials is mainly caused by the nontrivial transport behavior of phonons. They are the quantization of lattice vibration, which carries thermal energy. The scattering of phonons induces heat conduction in solids. The thermal conductivity of dielectric solids was proposed by Debye et al. [73, 74] in analogy with the heat conduction in ideal gas

$$\kappa = \frac{1}{3}\rho C_V v_s \lambda_R, \tag{1.12}$$

where $\lambda_R = v_s \tau_R$ is the MFP of R processes, namely the distance between successive collisions. The phonon MFPs of ordinary materials at room temperature are among 1–100 nm. Therefore, when the characteristic sizes of materials shrink and are comparable with the phonon MFPs, the boundary scattering of phonons will significantly reduce the effective MFPs, which results in reduction of effective thermal conductivity of nanomaterial according to Eq. 1.12. Low-dimensional materials such as CNT and graphene have phonon MFPs as long as several micrometers at room temperature. Thus their apparent thermal conductivities increase with characteristic lengths in a bigger size range [43–49].

In analogy to the radiative heat transport equation, Majumdar [75] established the equation of phonon radiative transfer (EPRT) to characterize phonon transport in thin films. If the thickness of a film is much larger than the phonon MFPs, the EPRT reduces to the ordinary Fourier's law. When the thickness is much lower than MFPs, the EPRT gives the Stefan-Boltzmann law for radiative heat transfer of black body. In the intermediate range, the effective thermal conductivity is predicted as

$$\frac{\kappa_{\text{eff}}}{\kappa_0} = \frac{1}{1 + \beta\frac{\lambda_R}{l}}, \tag{1.13}$$

where l is the film thickness, $\beta = 3/8$ for the in-plane thermal conductivity of a film, while $\beta = 4/8$ for the cross plane thermal conductivity [76, 77]. Equation 1.13 is also named as the gray model, since it assumes the same MFP for all phonons regardless of mode and frequency.

Chen [59] solved radiative BTE to characterize the reduction of effective thermal conductivity in the cross plane direction of GaAs/AlAs, Si/Ge, and Bi_2Te_3/Sb_2Te_3 superlattices. His main assumptions include: (1) The thickness of each layer of superlattices is much larger than the phonon wavelength, so as to solve the BTE with the phonon spectrum obtained for bulk material. (2) A single MFP is used to approximate the scattering of phonons in bulk material (i.e., a single relaxation time). The interfacial thermal resistance is determined by the reflection and transmission rates. Therefore, it is proposed that the MFP extracted from Eq. 1.12 is unable to predict the size effects. The reason is that the long wave length phonons contribute more to heat conduction while the optical phonons contribute more to the specific heat. Re-examining the issue with the assumed sinusoidal phonon dispersion relation leads to the modification of phonon MFPs. For the single crystal Si, the MFP based on Eq. 1.12 is 40.9 nm, while the modified one is 260 nm. Through the numerical solution of BTE, it is found that the temperature drop mainly happens at the interface in nanostructured superlattices due to the compatibility between the thickness of superlattice layers and phonon MFPs. Yang and Chen [60], Dames and Chen [61] further studied heat conduction in Si/Ge superlattice nanowires and two-dimensional Si/Ge nanocomposites. The effective thermal conductivities of these systems reduce with the characteristic size of each component, and are significantly affected by the reflection and transmission rates of interfaces.

Asheghi et al. [50] measured the in-plane thermal conductivity of single crystal Si thin films with thicknesses of 1.60, 0.83, and 0.42 μm, which were deposited on the SiO_2 substrate. They found that at room temperature the effective thermal conductivity is close to the bulk value for 1.60 μm sample, while it reduces slightly for 0.83 and 0.42 μm films. At low temperature (30 K), two orders of magnitude reduction from the bulk value was observed in these samples. They elucidated that if the thickness of Si nanofilms is lower than 50 nm, the effective thermal conductivity could reduce by 70 % from the bulk value at room temperature. Ju and Goodson [51] prepared single crystal Si thin films with thicknesses between 74 and 240 nm through repeatedly oxidizing Si films and then removing the oxidized layers. The in-plane thermal conductivities of these thin films were measured. Based on the measurement they proposed that the LA branch of phonons contribute mainly to the heat conduction. The relaxation time satisfies $\tau^{-1} \propto \omega^{1.7}$, with ω the phonon frequency. They estimated that the phonon MFP is 300 nm based on the phonon BTE, which is close to Chen et al.'s theoretical result. Liu and Asheghi [52], Ju [53] measured the thermal conductivities of thinner single crystal Si films. The trends is consistent to [51]. The experimental results for Si nanofilms are shown in Fig. 1.1a.

In 2003, Li et al. [54] measured the effective thermal conductivity of suspended single crystal Si nanowire produced by the vapor-liquid-solid (VLS) method (Fig. 1.1b). The results show that the effective thermal conductivities of nanowires are apparently lower than the bulk value and decrease with the reduction in diameters. The size effect also causes the shift of peaks of the thermal conductivity versus temperature curves for nanowires. The bulk material has the peak at around 25 K. The peaks for nanowires with diameters of 37, 56, and 115 nm are at 210, 160, 130 K, respectively. Such shift indicates that the phonon-boundary scattering strongly affects the thermal conductivity for thin nanowires. In the range of 20–60 K, the temperature dependence relations of thermal conductivity for 56 and 115 nm nanowires are approximately proportional to T^3. Li elucidated that in this condition the boundary scattering dominates, causing effective thermal conductivities proportional to the specific heat. However, for 37 and 22 nm nanowires, the

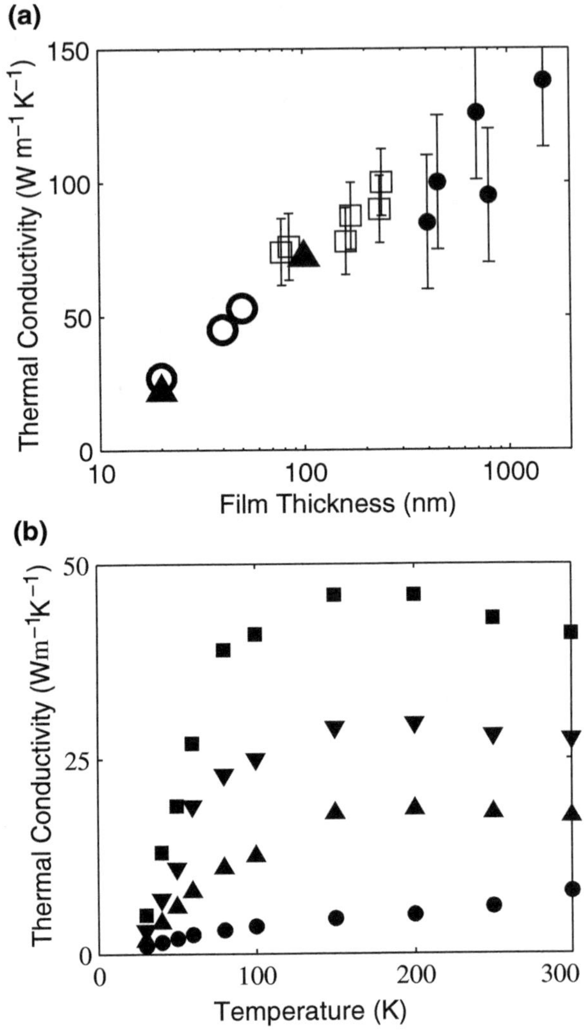

Fig. 1.1 Experimental results of nanofilms and nanowires. **a** In-plane effective thermal conductivity of single crystal Si nanofilms at 300 K, (*filled circle*) [50], (*open square*) [51], (*filled triangle*) [52], (*open circle*) [53]. **b** Effective thermal conductivity of single crystal Si nanowires with diameter of 115 nm (*filled square*), 56 nm (*filled down triangle*), 37 nm (*filled triangle*), 22 nm (*filled circle*) [54]

temperature dependences of thermal conductivity are close to $\propto T^2$ and $\propto T$, respectively. This effect was attributed to the reduction of phonon effective group velocity due to the change of phonon dispersion relation. Mingo [78] considered the complete dispersion relation of phonons and adopted Matheissen rule to average the relaxation times of the boundary scattering, Umklapp scattering, and impurity scattering. He obtained the theoretical prediction very close to Li et al.'s results. Nevertheless, it is notable that Mingo used nanowires of 38.5, 72.8, and 132.5 nm in the theoretical model to compare with the 37, 56, and 115 nm nanowires in the experimental results. Some arbitrariness thus exists in the assumed boundary slip parameter. On the other hand, the experimental results for 22 nm nanowire are poorly explained.

In 2008, Hochbaum et al. [55] reported the aqueous electroless etching (EE) synthesis of Si nanowires with boundary roughness between 1 and 5 nm. The nanowires obtained through the EE method have a similar temperature dependence trend of thermal conductivity to those synthesized by VLS method. However, the absolute value further decreases by five to eightfolds. This reduction is attributed to the secondary scattering of phonons induced by the rough boundary. The ultralow lattice thermal conductivity increases the thermoelectric figure of merit (ZT) of Si nanowires. The thermal conductivity of the boron doped Si nanowire of diameter 52 nm can reduce to 2 W/(m K) at room temperature, giving ZT as high as 0.6. Such low thermal conductivity cannot be explained by Mingo's theory. Even adding the boundary back scattering in his model is not enough to reproduce the experimental results. Si nanowires with high ZT have attractive potential and stimulate much of the following research. Hippalgaonkar et al. [79] made the rough nanowire with a rectangle cross section by the electron beam lithography (EBL) method. The characteristic size is around 80 nm. The effective thermal conductivity is lower than that of nanowires made by VLS method but it is still higher than those made by EE method. Moore et al. [80] carried out Monte Carlo (MC) simulation of sawtooth nanowires. The cross section is 22 nm × 22 nm square. The results show that the thermal conductivities of these nanowires strongly decrease from the bulk value, however, they are still considerably higher than the experimental results by Hochbaum et al., Carrete et al. [81], He and Galli [82], Sullivan and coworkers [83] studied nanowires with rough and oxidized boundary by MD simulation. Results show that the rough boundary can induce huge reduction of thermal conductivity, but not as large as Hochbaum's experiments. Thus the relation between boundary roughness and ultralow thermal conductivity is still unclear.

The above-mentioned steady non-Fourier heat conduction is caused by the size effect of nanomaterials. Wang et al. [84, 85] observed that in low temperature metal thin films, the large heat flux will cause deviation of Fourier's law. The mechanism of this effect is different from the size effect in dielectric nanosystems, and can be attributed to the inertia of thermomass.

The Fourier's law is a parabolic equation. It is valid when the energy carriers transport diffusively. In fast transient heating and nanosystems, the energy carriers (e.g., phonons) cannot fully relax to the near equilibrium state due to the limitation of time and spatial scales. Therefore, the transport is partial ballistic and leads to the failure of Fourier's law. In recent years, new measurement methods such as TDTR [86, 87], thermal grating [88] and frequency domain thermoreflectance [89] have been used to study the phonon spectrum and MFPs in dielectric materials. In these experiments the transient and steady non-Fourier heat conduction will coexist and the analysis is complex. The present theory relies mainly on the numerical solution of the microscopic BTE and Green's function to explain the experimental results. The solution process requires much assumption on parameters and induces arbitraries. Thus, a more general heat conduction model with explicit macroscopic physical meaning which is suitable for non-Fourier heat conduction is highly desired.

1.3 Non-Fourier Heat Conduction and Irreversible Thermodynamics

1.3.1 Classical Irreversible Thermodynamics

The classical thermodynamic theory was established in the nineteenth century, which is based on the equilibrium system. Neither the time needed for the change of system status nor the mass and energy transfer in nonuniform systems is considered. In 1931, Onsager [90] derived the reciprocal relation for irreversible processes based on the hypothesis of microscopic reversibility and the linear regression of fluctuation. The phenomenological coefficients of linear coupling among various irreversible processes are in symmetry according to the reciprocal relations. In this way, the thermodynamic theory is extended to nonequilibrium systems. The Onsager's derivation and assumption for irreversible processes was further developed by a series of works [90–104] that evolve into the foundation of modern nonequilibrium thermodynamics (also called thermodynamics of irreversible processes, TIP).

The entropy production rate, σ^s, is the key variable in nonequilibrium thermodynamics. It can be written as the bilinear product of generalized forces (thermodynamic forces), X, and generalized fluxes (thermodynamic fluxes), J

$$\sigma^s = \sum_{\alpha} \boldsymbol{J}_{\alpha} \boldsymbol{X}_{\alpha} \tag{1.14}$$

Equation 1.14 can be expressed for usual irreversible transport processes as

$$\sigma^s = \boldsymbol{q} \cdot \nabla\left(\frac{1}{T}\right) - \frac{1}{T}\mathbf{P}^v : \nabla \boldsymbol{u}_{\mathrm{f}} - \sum_{k=1}^{N} \boldsymbol{J}_k \cdot \nabla\left(\frac{\mu_k}{T}\right) + \frac{1}{T}\boldsymbol{i} \cdot \nabla \varphi_{\mathrm{e}}, \tag{1.15}$$

where $\mathbf{P}^v$ is the stress tensor, $\boldsymbol{u}_{\mathrm{f}}$ is the fluid velocity, $\boldsymbol{J}_k$ is the mass component diffusion flux, μ_k is the chemical potential, $\boldsymbol{i}$ is the electric flux density, and φ_{e} is the electric potential. The terms on the right-hand side of Eq. 1.15 denote the entropy production caused by the heat conduction, momentum diffusion, mass component diffusion, and electrical conduction, respectively. Assume a linear relation between generalized forces and fluxes, with the scaling factors and the phenomenological parameters one obtains the linear transport laws for each irreversible process

$$\boldsymbol{q} = -\kappa \nabla T \tag{1.16a}$$

$$\boldsymbol{J}_k = -\sum_{j=1}^{N} \rho D_{kj} \nabla c_j \tag{1.16b}$$

$$\boldsymbol{i} = -\frac{1}{r_{\mathrm{e}}}\nabla\varphi_{\mathrm{e}} \tag{1.16c}$$

$$\mathbf{P}^{v} = -2\eta\nabla\boldsymbol{u}_{\mathrm{f}}, \tag{1.16d}$$

where κ is the thermal conductivity, η is the viscosity, D is the diffusivity, $c_k = \rho_k/\rho$ is the mass concentration ratio, and r_{e} is the electric resistivity. The four equations in Eqs. 1.16a–1.16d correspond to the Fourier's heat conduction law, Fick's law, Ohm's law, and Newton's viscosity law. When multi-transport processes coexist in the system, these irreversible processes can couple with each other, generating effects such as thermoelectric, thermophoresis, electrophoresis. In this case, the linear transport laws should be expressed more generally as

$$\boldsymbol{q} = L_{qq}\nabla(1/T) - \sum_{k=1}^{N} L_{qk}\nabla(\mu_k/T) - L_{qe}\nabla(\varphi_{\mathrm{e}}/T) \tag{1.17a}$$

$$\boldsymbol{J}_k = L_{kq}\nabla(1/T) - \sum_{j=1}^{N} L_{kj}\nabla(\mu_j/T) - L_{ke}\nabla(\varphi_{\mathrm{e}}/T) \tag{1.17b}$$

$$\boldsymbol{i} = L_{eq}\nabla(1/T) - \sum_{k=1}^{N} L_{ek}\nabla(\mu_k/T) - L_{ee}\nabla(\varphi_{\mathrm{e}}/T) \tag{1.17c}$$

Equations 1.17a–1.17c can be simply written as $J = LX$, where L is a symmetry matrix according to Onsager reciprocal relation. Such reciprocity gives the constraint of phenomenological coefficients such as the second Thomson relation from the thermodynamic perspective. Having shown its value in many fields, the classical nonequilibrium thermodynamics still has some defects. For example, the decomposition of entropy production into generalized forces and fluxes is to some extent arbitrary, which could break the Onsager reciprocal relation [96]. Its derivation is mostly based on the linear transport laws. Thus in the case of nonlinear transport, e.g., non-Fourier heat conduction, its applicability is questionable. Therefore, more general thermodynamic theory is needed for the nonlinear irreversible processes.

1.3.2 Extended Irreversible Thermodynamics (EIT)

The classical nonequilibrium thermodynamic theory is challenged in the case of non-Fourier heat conduction. For example, one considers the pure heat conduction in rigid solids. Inserting the CV model, Eq. 1.4, into the entropy production, Eq. 1.15, yields

$$\sigma^s = \frac{\kappa \nabla T \cdot \nabla T}{T^2} + \frac{\tau}{T^2}\frac{\partial \boldsymbol{q}}{\partial t} \cdot \nabla T \tag{1.18}$$

Equation 1.18 is no longer quadratic. It cannot keep semi-positive definite. The negative entropy production implies that the wave-like heat conduction may violate the second law of thermodynamics, which is a paradox. The theory of extended irreversible thermodynamics (EIT) [105–120] proposes to introduce more state variables to eliminate this paradox. When the Fourier's law holds, the temperature gradient in the system is correlated to the heat flux, which means that they depend on each other. EIT raises the heat flux as independent state variables in addition to the internal energy (temperature). The local entropy density, s, is thus modified as $s = s(e, q)$, where e is the density of internal energy. Combined with the CV model one obtains the derivative of extend entropy

$$\mathrm{d}s_{\mathrm{EIT}} = T^{-1}\mathrm{d}e - \frac{\tau}{\rho\kappa T^2}\boldsymbol{q} \cdot \mathrm{d}\boldsymbol{q} \tag{1.19}$$

Making integral gives

$$s_{\mathrm{EIT}}(e, \boldsymbol{q}) = s_{\mathrm{eq}}(e) - \frac{1}{2}\frac{\tau}{\kappa T^2}\boldsymbol{q} \cdot \boldsymbol{q} \tag{1.20}$$

The corresponding extended entropy production is

$$\sigma^s_{\mathrm{EIT}} = \frac{1}{\kappa T^2}\boldsymbol{q} \cdot \boldsymbol{q} \tag{1.21}$$

In contrast to Eqs. 1.18 and 1.21 recovers the quadratic form and keeps semi-positive definite during heat wave propagation [109]. Figure 1.2 shows the time evolution of entropy in an isolated one-dimensional system with heat wave propagation. It can be seen that the classical entropy evolution is non-monotonic while the extended entropy keeps monotonic. Thus the second law of thermodynamics is recovered by the EIT.

Since the entropy and entropy production are modified in EIT, the thermodynamic temperature should also be modified. The EIT derives the nonequilibrium temperature, θ, based on the relation in classical thermodynamics $T^{-1} = \mathrm{d}s/\mathrm{d}e$ [107, 110, 111]

$$\theta = T_{\mathrm{eq}}\left(1 - \tau \middle/ \rho\kappa C_{\mathrm{V}} T^2_{\mathrm{eq}}\boldsymbol{q} \cdot \boldsymbol{q} + o\left(q^2\right)\right), \tag{1.22}$$

where T_{eq} is the temperature in equilibrium system. Equation 1.22 indicates that the nonequilibrium temperature will be lower than the equilibrium one when heat flux passes through the system.

EIT theory has been applied in the field of non-Fourier heat conduction, non-Fick diffusion as well as non-Newtonian momentum transport. A series of

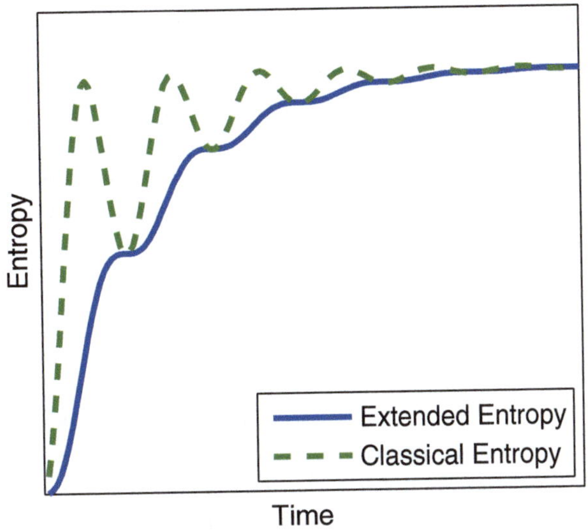

Fig. 1.2 Entropy evolution in an isolated 1D system with heat wave propagation [109]. At $t = 0$ the distribution of temperature is sinusoidal, then the evolution satisfies CV model. *Solid line* EIT entropy; *Dash line* classical entropy

nonlinear constitutive equations for nonlinear transports are derived based on EIT. For example, adding a transient relaxation term in the nonequilibrium temperature, θ, gives the dynamic nonequilibrium temperature [112–114], based on which the nonlinear heat conduction model with nonlocal terms can be derived. Such model can be used to describe the heat wave dispersion and thermal chocking. By further upgrading the gradient of heat flux, ∇q, as the independent system variable, EIT can give the phonon hydrodynamics model, Eq. 1.7, which characterizes the size effect of thermal conductivity in nanosystems [115–119].

EIT removes the paradox of negative entropy production in non-Fourier heat conduction. However, its derivation, to some extent, is from an ad hoc perspective. The decomposition of generalized forces and fluxes still lacks explicit physical definition. The nonequilibrium temperature is mathematically based on the extended entropy, and needs more discussion on its physical meaning. Therefore, nonequilibrium thermodynamics dealing with non-Fourier heat conduction is to be further developed from more ab initio perspectives.

1.4 Conclusion

The word thermodynamics originates from the Greek words θέρμη therme, meaning "heat," and δύναμις dynamis, meaning "power" [121]. The classical thermodynamics is not the "dynamics of heat," but instead a subject on the heat engine. It is only applicable in equilibrium systems and does not consider the time needed for state evolution. The nonequilibrium thermodynamics and the heat transfer theory have the concept of time and rate, but still lack true dynamic variables such as force and momentum. In recent years, Guo et al. [122] proposed that the heat has a dual nature of energy and mass. According to Einstein's mass energy equivalence, the thermal energy stored in the system has its equivalent mass, i.e., thermomass. Though tiny, the balance and motion of thermomass represents the heat transfer and can be analyzed by the dynamic framework. Based on this elucidation, Guo et al. developed the thermomass theory [25, 84, 85, 122–126] dealing with the non-Fourier heat conduction and the entransy theory [127–131] dealing with the optimization of heat transfer system. In this work, the non-Fourier heat conduction is investigated from the dynamic point of view based on thermomass. The following issues are addressed:

1. The present general heat conduction law derived from thermomass theory is macroscopic and phenomenological. It has a similar form to the phonon hydrodynamic model based on the phonon BTE. This work explores the microscopic foundation of the thermomass theory based on BTE, and compares it with the phonon hydrodynamics. It bridges the macroscopic and microscopic theories for non-Fourier heat conduction.
2. The irreversible thermodynamics is discussed from the dynamical viewpoint. The novel physical interpretation of thermodynamic quantities such as entropy and temperature is compared with EIT. The defects in the proof of Onsager reciprocal relation also are investigated with the present theory framework.
3. The non-Fourier heat conduction in nanosystems is analyzed with the dynamical theory. The novel models for the size-dependent thermal conductivity of nanosystems are developed and compared with the experiments.

References

1. Fourier J (1955) Analytical theory of heat. Dover Publications, New York
2. Cattaneo C (1948) Sulla conduzione del calore. Atti Sem Mat Fis Univ Modena 3:83–101
3. Vernotte P (1958) Paradoxes in the continuous theory of the heat equation. CR Acad Sci 246:3154–3155
4. Morse PM, Feshbach H (1953) Methods of theoretical physics. McGraw-Hill, New York
5. Hu RF, Cao BY (2009) Study on thermal wave based on the thermal mass theory. Sci China Series E Technol Sci 52(6):1786–1792
6. Körner C, Bergmann HW (1998) The physical defects of the hyperbolic heat conduction equation. Appl Phys A 67(4):397–401
7. Bai C, Lavine AS (1995) On hyperbolic heat conduction and the second law of thermodynamics. J Heat Transf 117(2):256–263
8. Ackerman CC, Bertman B, Fairbank HA, Guyer RA (1966) Second sound in solid helium. Phys Rev Lett 16:789
9. Rogers SJ (1971) Transport of heat and approach to second sound in some isotopically pure Alkali-Halide crystals. Phys Rev B 3(4):1440
10. Jackson HE, Walker CT (1971) Thermal conductivity, second sound, and phonon-phonon interactions in NaF. Phys Rev B 3(4):1428
11. Mason WP (1968) Physical acoustics: principles and methods, vol V. Academic Press, New York, pp 233–292
12. Ackerman CC (1968) Temperature pulses in dielectric solids. Ann Phys 50:128–185

13. Narayanamurti V, Dynes RC (1972) Observation of second sound in bismuth. Phys Rev Lett 28(22):1461
14. Joseph DD, Preziosi L (1989) Heat waves. Rev Mod Phys 61(1):41
15. Guyer RA, Krumhansl JA (1966) Solution of the linearized phonon Boltzmann equation. Phys Rev 148(2):766–778
16. Guyer RA, Krumhansl JA (1966) Thermal conductivity, second sound, and phonon hydrodynamic phenomena in nonmetallic crystals. Phys Rev 148(2):778
17. Eesley GL (1983) Observation of nonequilibrium electron heating in copper. Phys Rev Lett 51(23):2140–2143
18. Fujimoto JG, Ippen EP, Liu JM et al (1984) Femtosecond laser interaction with metallic tungsten and nonequilibrium electron and lattice temperatures. Phys Rev Lett 53:1837–1840
19. Eesley GL (1986) Generation of nonequilibrium electron and lattice temperatures in copper by picoseconds laser pulses. Phys Rev B 33(4):2144–2151
20. Schoenlein RW, Lin WZ, Fujimoto JG, Eesley GL (1987) Femtosecond studies of nonequilibrium electronic processes in metals. Phys Rev Lett 58(16):1680
21. Brorson SD, Fujimoto JG, Ippen EP (1987) Femtosecond electronic heat-transport dynamics in thin gold-films. Phys Rev Lett 59(17):1962–1965
22. Brorson SD, Kazeroonian A, Moodera JS et al (1990) Femtosecond room-temperature measurement of the electron-phonon coupling constant γ in metallic superconductors. Phys Rev Lett 64(18):2172
23. Sigel R, Tsakiris GD, Lavarenne F et al (1990) Experimental observation of laser-induced radiation heat waves. Phys Rev Lett 65(5):587
24. Wang JC, Guo CL (2007) Effect of electron heating on femtosecond laser-induced coherent acoustic phonons in noble metals. Phys Rev B 75:184304
25. Wang HD, Ma WG, Zhang X, Wang W (2010) Measurement of the thermal wave in metal films using femtosecond laser thermoreflectance system. Acta Phys Sin 59(6):3856–3862
26. Qiu TQ, Tien CL (1993) Heat transfer mechanisms during short-pulse laser heating of metals. J Heat Transf 115(4):835–841
27. Qiu TQ, Tien CL (1994) Femtosecond laser heating of multi-layer metals—I analysis. Int J Heat Mass Transf 37(17):2789–2797
28. Qiu TQ, Tien CL (1994) Femtosecond laser heating of multi-layer metals—II experiments. Int J Heat Mass Transf 37(17):2799–2808
29. Özişik MN, Tzou DY (1994) On the wave theory in heat conduction. J Heat Transf 116 (3):526–535
30. Tzou DY (1995) The generalized lagging response in small-scale and high-rate heating. Int J Heat Mass Transf 38(17):3231–3240
31. Tzou DY (1996) Macro-to microscale heat transfer: the lagging behavior. Taylor & Francis, Washington, DC
32. Antaki PJ (1998) Solution for non-Fourier dual phase lag heat conduction in a semi-infinite slab with surface heat flux. Int J Heat Mass Transf 41:2253–2258
33. Tang DW, Araki N (1999) Wavy, wavelike, diffusive thermal responses of finite rigid slabs to high-speed heating of laser-pulses. Int J Heat Mass Transf 42:855–860
34. Xu M, Wang L (2002) Thermal oscillation and resonance in dual-phase-lagging heat conduction. Int J Heat Mass Transf 45(5):1055–1061
35. Xu M, Wang L (2005) Dual-phase-lagging heat conduction based on Boltzmann transport equation. Int J Heat Mass Transf 48(25):5616–5624
36. Shen B, Zhang P (2008) Notable physical anomalies manifested in non-Fourier heat conduction under the dual-phase-lag model. Int J Heat Mass Transf 51(7):1713–1727
37. Chen G (2001) Ballistic-diffusive heat-conduction equations. Phys Rev Lett 86(11):2297–2300
38. Christov CI, Jordan PM (2005) Heat conduction paradox involving second-sound propagation in moving media. Phys Rev Lett 94:154301
39. Müller I, Ruggeri T (1993) Extended thermodynamics. Springer, New York

40. Tsai DH, MacDonald RA (1976) Molecular-dynamical study of second sound in a solid excited by a strong heat pulse. Phys Rev B 14(10):4714
41. Volz S, Saulnier JB, Lallemand M et al (1996) Transient Fourier-law deviation by molecular dynamics in solid argon. Phys Rev B 54(1):340
42. Shiomi J, Maruyama S (2006) Non-Fourier heat conduction in a single-walled carbon nanotube: classical molecular dynamics simulations. Phys Rev B 73(20):205420
43. Hone J, Whitney M, Piskoti C et al (1999) Thermal conductivity of single-walled carbon nanotubes. Phys Rev B 59(4):R2514
44. Berber S, Kwon YK, Tomanek D (2000) Unusually high thermal conductivity of carbon nanotubes. Phys Rev Lett 84(20):4613
45. Kim P, Shi L, Majumdar A et al (2001) Thermal transport measurements of individual multiwalled nanotubes. Phys Rev Lett 87(21):215502
46. Pop E, Mann D, Wang Q et al (2006) Thermal conductance of an individual single-wall carbon nanotube above room temperature. Nano Lett 6(1):96–100
47. Ghosh S, Calizo I, Teweldebrhan D et al (2008) Extremely high thermal conductivity of graphene: prospects for thermal management applications in nanoelectronic circuits. Appl Phys Lett 92(15):151911
48. Balandin AA, Ghosh S, Bao W et al (2008) Superior thermal conductivity of single-layer graphene. Nano Lett 8(3):902–907
49. Guo Z, Zhang D, Gong XG (2009) Thermal conductivity of graphene nanoribbons. Appl Phys Lett 95(16):163103
50. Asheghi M, Leung YK, Wong SS et al (1997) Phonon-boundary scattering in thin silicon layers. Appl Phys Lett 71(13):1798–1800
51. Ju YS, Goodson KE (1999) Phonon scattering in silicon films with thickness of order 100 nm. Appl Phys Lett 74(20):3005–3007
52. Liu W, Asheghi M (2004) Phonon–boundary scattering in ultrathin single-crystal silicon layers. Appl Phys Lett 84(19):3819–3821
53. Ju YS (2005) Phonon heat transport in silicon nanostructures. Appl Phys Lett 87(15):153106
54. Li D, Wu Y, Kim P et al (2003) Thermal conductivity of individual silicon nanowires. Appl Phys Lett 83(14):2934–2936
55. Hochbaum AI, Chen R, Delgado RD et al (2008) Enhanced thermoelectric performance of rough silicon nanowires. Nature 451(7175):163–167
56. Boukai AI, Bunimovich Y, Tahir-Kheli J et al (2008) Silicon nanowires as efficient thermoelectric materials. Nature 451(7175):168–171
57. Lee SM, Cahill DG, Venkatasubramanian R (1997) Thermal conductivity of Si–Ge superlattices. Appl Phys Lett 70(22):2957–2959
58. Chen G, Neagu M (1997) Thermal conductivity and heat transfer in superlattices. Appl Phys Lett 71(19):2761–2763
59. Chen G (1998) Thermal conductivity and ballistic-phonon transport in the cross-plane direction of superlattices. Phys Rev B 57(23):14958
60. Yang R, Chen G (2004) Thermal conductivity modeling of periodic two-dimensional nanocomposites. Phys Rev B 69(19):195316
61. Dames C, Chen G (2003) Theoretical phonon thermal conductivity of Si/Ge superlattice nanowires. J Appl Phys 95(2):682–693
62. Capinski WS, Maris HJ, Ruf T et al (1999) Thermal-conductivity measurements of GaAs/AlAs superlattices using a picosecond optical pump-and-probe technique. Phys Rev B 59(12):8105
63. Li D, Wu Y, Fan R et al (2003) Thermal conductivity of Si/SiGe superlattice nanowires. Appl Phys Lett 83(15):3186–3188
64. Luckyanova MN, Garg J, Esfarjani K et al (2012) Coherent phonon heat conduction in superlattices. Science 338(6109):936–939
65. Henry A, Chen G (2008) High thermal conductivity of single polyethylene chains using molecular dynamics simulations. Phys Rev Lett 101(23):235502

66. Shen S, Henry A, Tong J et al (2010) Polyethylene nanofibres with very high thermal conductivities. Nat Nanotechnol 5(4):251–255
67. Cao BY, Li YW, Kong J et al (2011) High thermal conductivity of polyethylene nanowire arrays fabricated by an improved nanoporous template wetting technique. Polymer 52 (8):1711–1715
68. Cahill DG, Ford WK, Goodson KE et al (2002) Nanoscale thermal transport. J Appl Phys 93 (2):793–818
69. Kim W, Wang R, Majumdar A (2007) Nanostructuring expands thermal limits. Nano Today 2(1):40–47
70. Chen G (2005) Nanoscale energy transport and conversion: a parallel treatment of electrons, molecules, phonons, and photons. Oxford University Press, Oxford
71. Baxter J, Bian Z, Chen G et al (2009) Nanoscale design to enable the revolution in renewable energy. Energy Environ Sci 2(6):559–588
72. Minnich A, Dresselhaus MS, Ren ZF, Chen G (2009) Bulk nanostructured thermoelectric materials: current research and future prospects. Energy Environ Sci 2(5):466–479
73. Kittel C (1996) Introduction to solid state physics, 7th edn. Wiley, New York
74. Reissland JA (1973) The physics of phonons. Wiley, London
75. Majumdar A (1993) Microscale heat-conduction in dielectric thin-films. J Heat Transfer 115 (1):7–16
76. McGaughey AJH, Landry ES, Sellan DP et al (2011) Size-dependent model for thin film and nanowire thermal conductivity. Appl Phys Lett 99(13):131904
77. Ziman JM (2001) Electrons and phonons: the theory of transport phenomena in solids. Oxford University Press, Oxford
78. Mingo N (2003) Calculation of Si nanowire thermal conductivity using complete phonon dispersion relations. Phys Rev B 68(11):113308
79. Hippalgaonkar K, Huang B, Chen R et al (2010) Fabrication of microdevices with integrated nanowires for investigating low-dimensional phonon transport. Nano Lett 10(11):4341–4348
80. Moore AL, Saha SK, Prasher RS et al (2008) Phonon backscattering and thermal conductivity suppression in sawtooth nanowires. Appl Phys Lett 93(8):083112
81. Carrete J, Gallego LJ, Varela LM, Mingo N (2011) Surface roughness and thermal conductivity of semiconductor nanowires: going below the Casimir limit. Phys Rev B 84 (7):075403
82. He Y, Galli G (2012) Microscopic origin of the reduced thermal conductivity of silicon nanowires. Phys Rev Lett 108(21):215901
83. Sullivan SE, Lin KH, Avdoshenko S et al (2013) Limit for thermal transport reduction in Si nanowires with nanoengineered corrugations. Appl Phys Lett 103(24):243107
84. Wang H, Liu J, Guo ZY, Takahashi K (2012) Non-Fourier heat conduction study for steady states in metallic nanofilms. Chin Sci Bull 57(24):3239–3243
85. Wang HD (2011) Theoretical and experimental studies on non-Fourier heat conduction based on thermomass theory. Doctoral dissertation, Tsinghua University, Beijing
86. Minnich AJ, Johnson JA, Schmidt AJ et al (2011) Thermal conductivity spectroscopy technique to measure phonon mean free paths. Phys Rev Lett 107(9):095901
87. Minnich AJ (2012) Determining phonon mean free paths from observations of quasiballistic thermal transport. Phys Rev Lett 109(20):205901
88. Johnson JA, Maznev AA, Cuffe J et al (2013) Direct measurement of room-temperature nondiffusive thermal transport over micron distances in a silicon membrane. Phys Rev Lett 110(2):025901
89. Regner KT, Sellan DP, Su Z et al (2013) Broadband phonon mean free path contributions to thermal conductivity measured using frequency domain thermoreflectance. Nat Commun 4:1640
90. Onsager L (1931) Reciprocal relations in irreversible processes I. Phys Rev 37(4):405
91. Onsager L (1931) Reciprocal relations in irreversible processes II. Phys Rev 38(12):2265
92. Onsager L, Machlup S (1953) Fluctuations and irreversible processes. Phys Rev 91(6): 1505–1512

93. Machlup S, Onsager L (1953) Fluctuations and irreversible processes. Phys Rev 91:1512–1515
94. Callen HB (1948) The application of Onsager's reciprocal relations to thermoelectric, thermomagnetic, and galvanomagnetic effects. Phys Rev 3(11):1349
95. Casimir HBG (1945) On Onsager's principle of microscopic reversibility. Rev Mod Phys 17 (2–3):343
96. Coleman BD, Truesdell C (1960) On the reciprocal relations of Onsager. J Chem Phys 33 (1):28–31
97. Gyarmati I, Gyarmati E (1970) Non-equilibrium thermodynamics. Springer, Berlin
98. Mazur P, de Groot SR (1963) Non-equilibrium thermodynamics. North-Holland, Amsterdam
99. Zeng DL (1991) Engineering non-equilibrium thermodynamics. Science Press, Beijing
100. Spera FJ, Trial AF (1993) Verification of the Onsager reciprocal relations in a molten silicate solution. Science 259(5092):204–206
101. Gallavotti G (1996) Chaotic hypothesis: Onsager reciprocity and fluctuation-dissipation theorem. J Stat Phys 84(5–6):899–925
102. Gabrielli D, Jona-Lasinio G, Landim C (1996) Onsager reciprocity relations without microscopic reversibility. Phys Rev Lett 77(7):1202–1205
103. Sharipov F (2006) Onsager-Casimir reciprocal relations based on the Boltzmann equation and gas-surface interaction: single gas. Phys Rev E 73(2):026110
104. Kocherginsky N, Gruebele M (2013) A thermodynamic derivation of the reciprocal relations. J Chem Phys 138(12):124502
105. Jou D, Casas-Vazquez J, Lebon G (1999) Extended irreversible thermodynamics revisited (1988–98). Rep Prog Phys 62(7):1035
106. Lebon G, Jou D, Casas-Vázquez J (2008) Understanding non-equilibrium thermodynamics. Springer, Berlin
107. Jou D, Casas-Vázquez J, Lebon G (2010) Extended irreversible thermodynamics. Springer, Berlin
108. Cimmelli VA (2009) Different thermodynamic theories and different heat conduction laws. J Non Equilibr Thermodyn 34(4):299–333
109. Criado-Sancho M, Llebot JE (1993) Behavior of entropy in hyperbolic heat conduction. Phys Rev E 47:4104–4107
110. Casas-Vázquez J, Jou D (1994) Nonequilibrium temperature versus local-equilibrium temperature. Phys Rev E 49(2):1040
111. Casas-Vazquez J, Jou D (2003) Temperature in non-equilibrium states: a review of open problems and current proposals. Rep Prog Phys 66(11):1937
112. Cimmelli VA, Sellitto A, Jou D (2009) Nonlocal effects and second sound in a nonequilibrium steady state. Phys Rev B 79(1):014303
113. Cimmelli VA, Sellitto A, Jou D (2010) Nonequilibrium temperatures, heat waves, and nonlinear heat transport equations. Phys Rev B 81(5):054301
114. Cimmelli VA, Sellitto A, Jou D (2010) Nonlinear evolution and stability of the heat flow in nanosystems: beyond linear phonon hydrodynamics. Phys Rev B 82(18):184302
115. Alvarez FX, Jou D (2008) Size and frequency dependence of effective thermal conductivity in nanosystems. J Appl Phys 103(9):094321
116. Sellitto A, Alvarez FX, Jou D (2010) Second law of thermodynamics and phonon-boundary conditions in nanowires. J Appl Phys 107(6):064302
117. Jou D, Criado-Sancho M, Casas-Vázquez J (2010) Heat fluctuations and phonon hydrodynamics in nanowires. J Appl Phys 107(8):084302
118. Jou D, Cimmelli VA, Sellitto A (2012) Nonlocal heat transport with phonons and electrons: application to metallic nanowires. Int J Heat Mass Transf 55(9):2338–2344
119. Sellitto A, Alvarez FX, Jou D (2012) Geometrical dependence of thermal conductivity in elliptical and rectangular nanowires. Int J Heat Mass Transf 5(11):3114–3120
120. Sellitto A, Cimmelli VA, Jou D (2013) Entropy flux and anomalous axial heat transport at the nanoscale. Phys Rev B 87(5):054302

121. Wikipedia (2014) Thermodynamics. http://en.wikipedia.org/wiki/Thermodynamics#cite_note-41. Accessed 08 Apr 2014
122. Guo ZY, Cao BY, Zhu HY et al (2007) State equation of phonon gas and conservation equations for phonon gas motion. Acta Phys Sin 56(6):3306–3312
123. Guo ZY, Cao BY (2008) A general heat conduction law based on the concept of motion of thermal mass. Acta Phys Sin 57(7):4273–4281
124. Cao BY, Guo ZY (2007) Equation of motion of a phonon gas and non-Fourier heat conduction. J Appl Phys 102(5):053503
125. Tzou DY, Guo ZY (2010) Nonlocal behavior in thermal lagging. Int J Therm Sci 49 (7):1133–1137
126. Guo ZY, Hou QW (2010) Thermal wave based on the thermomass model. J Heat Transf 132 (7):072403
127. Guo ZY, Zhu HY, Liang XG (2007) Entransy—a physical quantity describing heat transfer ability. Int J Heat Mass Transf 50(13):2545–2556
128. Zhu H Y (2007) The minimum thermal resistance principle based on entransy dissipation. Doctoral dissertation, Tsinghua University, Beijing
129. Chen Q (2008) Irreversibility and optimization of convective transport processes. Doctoral dissertation, Tsinghua University, Beijing
130. Chen Q, Liang XG, Guo ZY (2013) Entransy theory for the optimization of heat transfer—a review and update. Int J Heat Mass Transf 63:65–81
131. Cheng XT, Liang XG (2013) From thermomass to entransy. Int J Heat Mass Transf 62: 174–177

Chapter 2
Dynamical Governing Equations of Non-Fourier Heat Conduction

Abstract Thermal energy has its corresponding equivalent mass according to Einstein's mass–energy equivalence, which is termed as thermomass. The thermomass theory established the continuous governing equation for the non-Fourier heat conduction. The mass balance equation of thermomass gives the energy conservation equation while the momentum balance equation of thermomass gives the general heat conduction law. The microscopic foundation of the general heat conduction law based on the thermomass theory is investigated. The derivation based on the phonon Boltzmann equation shows that the second order expansion of phonon distribution function leads to the spatial inertia (or convective) term in the general heat conduction law, which makes the difference from the previous phonon hydrodynamic model. Limiting to the first order expansion will give the Cattaneo-Vernotte model, while the zeroth order expansion gives the classical Fourier's law. Comparison with other derivations of phonon Boltzmann equation such as the Chapman–Enskog expansion and eigenvalue analysis is presented.

2.1 Mass–Energy Duality of Heat

In history, the nature of heat is regarded as either a fluid (caloric theory) or a type of motion (dynamic theory, kinetic theory, or mechanical theory). The caloric theory regards heat as a weightless, self-repulsive fluid. In the eighteenth and the first half of nineteenth centuries, the caloric theory was the mainstream theory. It gave explanations to the temperature change and phase change, and even contributed to the establishment of the Fourier's heat conduction law and the Carnot's rule for thermal engines. However, in the nineteenth century, a series of experiments showed the conversion between heat and work, such as Rumford's report of boring a cannon and Joule's measurement of the mechanical equivalent of heat. After the mid-nineteenth century, the caloric theory was generally superseded by the mechanical theory along with the acceptance of energy conservation law and kinetic theory of gas. The modern thermodynamic and heat transport theory is

Y. Dong, *Dynamical Analysis of Non-Fourier Heat Conduction and Its Application in Nanosystems*, Springer Theses,
DOI 10.1007/978-3-662-48485-2_2

based on the understanding that heat originates from the individually random, or disordered, motion of particles in a body.

In the twentieth century, Einstein's relativity theory introduced the well-known mass–energy equivalence relation, $E = mc^2$, where c is the speed of light. Therefore, all forms of energy correspond to a certain amount of mass. Consider a system consisting n free particles, each having a mass m and a velocity v_i; the rest mass of this system based on the relativity theory is

$$m_{0,\mathrm{sys}} = \sum_{i=1}^{n} \frac{E_i}{c^2} \approx \sum_{i=1}^{n} \frac{m_{0,i}c^2 + (1/2)m_{0,i}v_i^2}{c^2} + o\left(v_i^2/c^2\right) \quad \text{with } \sum\nolimits_i m_{0,i}v_i = 0, \tag{2.1}$$

where m_0 is the rest mass of each particle. Equation 2.1 denotes that the rest mass of the system contains the total kinetic energy of individual particles, divided by c^2. Heating the system increases the energy of the disordered motion of particles while the system's net momentum preserves zero. The increased rest mass due to heating has the same nature as ordinary mass, like inertia. Thus, greater force is needed to propel a heated body with the same acceleration than when it is cool [1–4]. According to the general relativity, the inertia mass equals the gravitational mass. Thus "a piece of iron weighs more when red-hot than when cool" [5, 6]. The mass increase induced by heat, "thermomass," is very small in ordinary conditions. For example, the thermomass of Si at room temperature is 10^{-12} of the total mass. Such small amount of mass is negligible when dealing with dynamic problems such as movement and balance of the body. However, heat conduction is the movement of thermomass itself relative to the molecular or the lattice; so its mass, or inertia, should be accounted to describe the transport and motion of heat.

Moller [1] and Rindler et al. [4] analyzed the continuum mechanics in the relativity framework. The momentum flux for the relativity system with heat flux inside is expressed as

$$\boldsymbol{g} = \frac{h\boldsymbol{v} + \boldsymbol{v} \cdot \boldsymbol{\Pi} + \boldsymbol{J}}{c^2}, \tag{2.2}$$

where h is the relative energy density, v is the velocity of the body, $\boldsymbol{\Pi}$ is the relative stress tensor, and $\boldsymbol{J}$ is the energy flux density induced by heat conduction

$$\boldsymbol{J} = \gamma\left(\boldsymbol{q}^0 - \boldsymbol{v}\frac{\boldsymbol{q}^0 \cdot \boldsymbol{v}}{c^2}\frac{\gamma}{\gamma + 1}\right), \tag{2.3}$$

where γ is the Lorentz factor, $\gamma = (1 - \boldsymbol{v}^2/c^2)^{-1/2}$. $\boldsymbol{q}^0$ is the heat flux density in the rest reference framework. In low speed systems where v is much less than c, the difference between J and q is the small amount of the order of $(v/c)^2$. In this case Eq. 2.2 can be converted into

$$\boldsymbol{g}^0 = \left(\rho^0 + \rho_\mathrm{E}^0\right)\boldsymbol{v}^0 + \frac{\boldsymbol{v}^0 \cdot \boldsymbol{\Pi}^0}{c^2} + \frac{\boldsymbol{q}^0}{c^2}, \tag{2.4}$$

where the superscript 0 means the value in low speed system, ρ^0 is the density of rest mass of the molecular contained in the system, $\rho_\mathrm{E}{}^0$ is the density of the equivalent mass of the energy (e.g., elastic energy, chemical energy, thermal energy, and nuclear bonding energy) stored in the system, divided by c^2. Equation 2.4 shows that the momentum flux density of low speed system contains the momentum density of the rest mass (the first term on the right-hand side) plus the energy flux density divided by c^2 (the second and third terms on the right-hand side, which correspond to the work done by the stress and the heat flux, respectively). Note that the rest mass of the system contains the part contributed by the energy, $\rho_\mathrm{E}{}^0 v^0$. In normal conditions, the contribution by the heat flux is far more less than the first term. Nevertheless, consider a body at rest with heat conduction inside, i.e., $\boldsymbol{v}^0 = 0$, the heat flux term is the only nonzero one in Eq. 2.4.

Based on Eq. 2.4, in low speed conditions, the balance equation of mass and momentum can be expressed as

$$\frac{\mathrm{d}}{\mathrm{d}t^0}\left(\rho^0 + \rho_\mathrm{E}^0\right) + \nabla \cdot \boldsymbol{g}^0 = 0 \tag{2.5}$$

$$\frac{\mathrm{d}}{\mathrm{d}t^0}\boldsymbol{g}^0 + \nabla \cdot \boldsymbol{\Pi}^0 + \frac{\boldsymbol{q}^0 \cdot \nabla \boldsymbol{v}^0}{c^2} = 0 \tag{2.6}$$

Using the mass conservation relation

$$\frac{\mathrm{d}}{\mathrm{d}t^0}\rho^0 + \nabla \cdot \left(\rho^0 \boldsymbol{v}^0\right) = 0 \tag{2.7}$$

Equation 2.5 turns to

$$\frac{\mathrm{d}}{\mathrm{d}t^0}\rho_\mathrm{E}^0 + \nabla \cdot \left[\rho_\mathrm{E}^0 \boldsymbol{v}^0 + \frac{\boldsymbol{v}^0 \cdot \boldsymbol{\Pi}^0}{c^2} + \frac{\boldsymbol{q}^0}{c^2}\right] = 0 \tag{2.8}$$

If the velocity of the body is zero, then Eq. 2.8 reduces to the energy balance equation (cf. Eq. 1.2). If the velocity is nonzero, then Eq. 2.8 is actually the energy balance equation in convective conditions. Equation 2.6 is equivalent to the Navier–Stokes equation without the body forces if the third term is neglected. On the other hand, for a rest body with heat conduction, $v^0 = 0$, $\boldsymbol{q}^0 \neq 0$, the nonzero terms in Eq. 2.6 indicates that the heat flux is also driven by a stress tensor, as the momentum flux of rest mass. Note that Eq. 2.6 is built in the low speed system, thus the motion of q^0 obeys Newton's law of motion. In this way, for the heat flux in a rest body one has

$$\frac{\mathrm{d}}{\mathrm{d}t^0}\frac{\boldsymbol{q}^0}{c^2} = -\nabla \cdot \boldsymbol{\Pi}^0 \tag{2.9}$$

In ordinary conditions, the stress tensor, Π^0, acting on the heat flux is very small. However, the quantity of thermomass is also very small, so the thermomass can be driven by the stress tensor with a recognizable acceleration.

Based on the above discussion, the thermomass, although derived from the relativity theory, obeys Newton's law of motion as long as it is discussed in low speed systems. Therefore, the dynamical governing equation of the motion of thermomass can be established to analyze the heat conduction processes from the ab initio viewpoint.

2.2 Governing Equations of Phonon Gas Dynamics

Phonons are the main heat carriers of dielectric solids, which come from the quantization of the lattice vibration. The heat conduction in dielectric solids is thus determined by the transport of phonons. In the thermal equilibrium system, the distribution function of phonons, f, satisfies the Bose–Einstein distribution, namely obeys the Planck's law [7, 8]

$$f_0 = \left[\exp(\hbar\omega/k_\mathrm{B}T) - 1\right]^{-1}, \tag{2.10}$$

where $\hbar$ is the reduced Planck constant (Dirac constant), ω is the phonon frequency, and k_B is the Boltzmann constant. The collective behavior of phonons is like a gas, namely phonon gas. The energy carried by the phonon gas in a unit volume is

$$\begin{aligned} e = \rho C_V T &= (2\pi)^{-3}\sum_n \int [\hbar\omega^n f^n(\boldsymbol{k},\boldsymbol{x},t)]\mathrm{d}^3k \\ &= \sum_n \int_{\boldsymbol{k}} \hbar\omega^n f^n(\boldsymbol{k},\boldsymbol{x},t) \end{aligned}, \tag{2.11}$$

where i is the number of phonon branches, k is the wave vector. Inserting Eq. 2.11 into Eq. 2.1 gives

$$m_{0,\mathrm{sys}} = \frac{\sum_i m_{0,i}c^2 + \sum_n \int_k \hbar\omega^n f^n \mathrm{d}k}{c^2} + o\left(\sum_n \int_k \hbar\omega^n f^n \mathrm{d}k \Big/ c^2\right) \tag{2.12}$$

Thus the density of the equivalent mass of phonon gas can be defined as

$$\rho_\mathrm{h} = \frac{\rho C_V T}{c^2} \tag{2.13}$$

The motion of phonon gas is regarded as a weighable fluid flowing diffusively through a porous medium (the framework of lattice). The drift velocity is the mass flux q/c^2 dividide by the density ρ_h

$$\boldsymbol{u}_\mathrm{h} = \frac{\boldsymbol{q}}{\rho C_V T} \tag{2.14}$$

With the definition of mass and velocity, one can establish the mass and momentum balance equations of the phonon gas

$$\frac{\partial \rho_\mathrm{h}}{\partial t} + \nabla \cdot (\rho_\mathrm{h} \boldsymbol{u}_\mathrm{h}) = 0 \tag{2.15}$$

$$\rho_\mathrm{h} \frac{\partial \boldsymbol{u}_\mathrm{h}}{\partial t} + (\rho_\mathrm{h} \boldsymbol{u}_\mathrm{h} \cdot \nabla) \boldsymbol{u}_\mathrm{h} + \nabla p_\mathrm{h} = \boldsymbol{f}_\mathrm{h}, \tag{2.16}$$

where p_h is the phonon gas pressure, f_h is the friction force impeding the phonon gas. Inserting Eqs. 2.13 and 2.14 into Eq. 2.15 yields the energy balance equation for solids, Eq. 1.2. The momentum balance equation of phonon gas describes the heat transport in dielectric solids. Guo et al. derived the expression for the phonon gas pressure [9–13]

$$p_\mathrm{h} = \gamma_\mathrm{G} \rho_\mathrm{h} C_V T = \frac{\gamma_\mathrm{G} \rho (C_V T)^2}{c^2}, \tag{2.17}$$

where γ_G is the Grüneisen parameter. In bulk materials, the friction force acting on the phonon gas is proportional to its drift velocity, which is similar to the case of porous flow

$$\boldsymbol{f}_\mathrm{h} = -\chi \rho_\mathrm{h} \boldsymbol{u}_\mathrm{h} \tag{2.18}$$

Equation 2.16 reduces to the Fourier's heat conduction law when the inertia effect is negligible. The friction factor in Eq. 2.18 can thereby be determined as

$$\chi = 2\gamma_\mathrm{G} \rho C_V^2 T / \kappa \tag{2.19}$$

Substitution of Eqs. 2.18 and 2.19 into the momentum balance relation, Eq. 2.16, gives the general heat conduction law [9–13]

$$\tau_\mathrm{TM} \frac{\partial \boldsymbol{q}}{\partial t} + 2(\boldsymbol{l} \cdot \nabla)\boldsymbol{q} - b\kappa \nabla T + \kappa \nabla T + \boldsymbol{q} = 0, \tag{2.20}$$

where

$$\tau_{\mathrm{TM}} = \frac{\kappa}{2\gamma_{\mathrm{G}}\rho C_V^2 T} \tag{2.21a}$$

$$\boldsymbol{l} = \frac{\boldsymbol{q}\kappa}{2\gamma_{\mathrm{G}} C_V (\rho C_V T)^2} = \boldsymbol{u}_{\mathrm{h}}\tau_{\mathrm{TM}} \tag{2.21b}$$

$$b = \frac{q^2}{2\gamma_{\mathrm{G}}\rho^2 C_V^3 T^3} = Ma_{\mathrm{h}}^2 \tag{2.21c}$$

τ_{TM} is the characteristic lagging time between the temperature gradient and the heat flux, which relates to the wave-like heat transport predicted by the relaxational models. The characteristic length l measures the spatial inertia of thermomass, which predicts the size effect in nanowires and nanotubes [14–16]. $v_{\mathrm{s}} = (2\gamma_{\mathrm{G}}C_V T)^{1/2}$ is the propagation speed of disturbance in phonon gas. The Mach number of phonon gas, $Ma_{\mathrm{h}} = u_{\mathrm{h}}/v_{\mathrm{s}}$, represents the compressibility of phonon gas, which predicts the heat flow chocking in CNTs [17]. The first term in Eq. 2.20 is the transient inertia. The second and third terms are the spatial inertia. The forth term is the driving force and the last term is the friction force. If all the inertia terms are neglected, one recovers the Fourier's heat conduction law, which is in analogy with the Darcy's law in porous flow. If only the spatial inertia term is negligible, one retrieves the Cattaneo-Vernotte (CV) model from Eq. 2.20.

Define the objective derivative as

$$\mathrm{D}/\mathrm{D}t = \partial/\partial t + 2(u_{\mathrm{h}} \cdot \nabla) \tag{2.22}$$

Thus Eq. 2.20 can be rewritten as

$$\tau_{\mathrm{TM}}\frac{\mathrm{D}\boldsymbol{q}}{\mathrm{D}t} + \boldsymbol{q} = -\kappa(1-b)\nabla T \tag{2.23}$$

The main differences between the general heat conduction law based on the thermomass theory and CV model are: (1) The objective derivative $\mathrm{D}/\mathrm{D}t$ replaces the partial derivative, $\partial/\partial t$; (2) The driving term contains additional parameter 1-b. The first difference arises from accounting for the convective effective of phonon gas, namely the spatial inertia. Christov and Jordan [18] indicated that in a moving medium, the objective derivative should be selected as $\mathrm{D}/\mathrm{D}t = \partial/\partial t + \boldsymbol{v} \cdot \nabla$, in order to avoid the paradox of breaking Galileo invariance during heat wave propagation. Note that v is the moving speed of the medium where heat conduction occurs, rather than the drift velocity of phonon gas. Usually the drift velocity of phonon gas is far less than the speed of conduction medium. However, in ultrasmall medium, $\boldsymbol{l}$, namely the product of drift velocity and relaxation time, could be comparable with the characteristic size of system. In this case the convection effect of phonon gas needs to be considered. Müller and Ruggeri [19] proposed that the objective derivative in the case of heat conduction in gases should have the form $\mathrm{D}q_i/\mathrm{D}t = \partial q_i/\partial t + q_k(\partial u_i/\partial x_k) - 2q_k W_{ik}$, where W is the angular velocity matrix.

The second term, $q_k(\partial u_i/\partial x_k)$, arises from the requirement of "objectivity", like the Jaumann derivative in the theory of non-Newtonian fluid. The third term can be written as $-2c^2(\rho_h u_h)_k W_{ik}$. It is actually the Coriolis effect induced by the rotation of conduction medium. The second difference relates to the definition of temperature during non-Fourier heat conduction, which will be discussed in detail in Chap. 4.

The derivation of Eq. 2.20 assumes the linear relation between friction and drift velocity. In the porous flow, when the flow region is near boundary, the Darcy's law needs to be modified into Darcy–Brinkman law. Thus the friction contains both the linear term of velocity and the second spatial derivative of velocity, i.e.,

$$\boldsymbol{f}_\mathrm{h} = -\chi\rho_\mathrm{h}\boldsymbol{u}_\mathrm{h} + \mu_\mathrm{h}\nabla^2\boldsymbol{u}_\mathrm{h} \tag{2.24}$$

where μ_h is the viscosity of phonon gas. Substitution of Eq. 2.24 into the momentum balance equation of phonon gas, Eq. 2.16, one has

$$\tau_\mathrm{TM}\frac{\partial \boldsymbol{q}}{\partial t} + 2l\nabla\boldsymbol{q} - b\kappa\nabla T + \kappa\nabla T + \boldsymbol{q} - l_\mathrm{B}^2\nabla^2\boldsymbol{q} = 0, \tag{2.25}$$

where

$$l_\mathrm{B} = \sqrt{\mu_\mathrm{h}/\chi\rho_\mathrm{h}} \tag{2.26}$$

The Brinkman term, $\mu_\mathrm{h}\nabla^2 u_\mathrm{h}$, indicates a boundary layer where the boundary friction is important. The characteristic thickness of the boundary layer for the phonon gas is l_B. At room temperature, l_B usually has the value 10 $\sim$ 100 nm. Therefore, the Brinkman effect only needs to be considered in nanosystems.

Equation 2.25 has a form similar to the phonon hydrodynamic model, Eq. 1.7. The latter is obtained through the solution of the linear phonon Boltzmann equation. In the following section the phonon Boltzmann derivation of Eq. 2.25 is discussed.

2.3 Microscopic Foundation

2.3.1 Phonon Boltzmann Derivation

As long as the system size is much larger than the phonon wavelength (typically less than a few nanometers for dielectric materials at room temperature), the aggregate behavior of phonons can be characterized by the Boltzmann equation. The state distribution function, *f*, is desired to be solved from the Boltzmann equation so as to give the governing equations of macroscopic quantities such as temperature and heat flux. Many assumptions are made in solving the phonon Boltzmann equation, which thereby influence the results of solution. The derivation here will be combined with the recognition that the phonon gas is a weighable fluid, and compared with other solution methods of the phonon Boltzmann equation.

The phonon Boltzmann equation generally has the form [20, 21]

$$\boldsymbol{D}f(\boldsymbol{k},\boldsymbol{x},t) = \boldsymbol{C}f(\boldsymbol{k},\boldsymbol{x},t) \tag{2.27}$$

where D and C are the drift and collision operator, respectively. The macroscopic variables, such as the internal energy density e and the heat flux density q, can be obtained by the integral of microscopic distribution function

$$e = \sum_n \int_{\boldsymbol{k}} \hbar\omega^n f^n \tag{2.28}$$

$$q_i = \sum_n \int_{\boldsymbol{k}} \hbar\omega^n \frac{\partial \omega^n}{\partial k_i} f^n \tag{2.29}$$

Guyer and Krumhansl [20, 21] obtained the eigenvalue solution of Eq. 2.27 which leads to the phonon hydrodynamics model, Eq. 1.7. Sussmann and Thellung [22] also obtained the governing equation similar to Eq. 1.7 by assuming that the Umklapp scattering rate is rare in pure crystals at low temperature. The Umklapp scattering and other phonon quasi-momentum ($\hbar k$) breaking scattering processes are called the resistive (R) processes. In contrast, the normal (N) scattering processes conserve the phonon quasi-momentum. The R processes incline to relax f to the equilibrium Planck distribution, i.e., f_0 in Eq. 2.10. The N processes incline to relax f to the displaced Planck distribution

$$f_{\mathrm{D}} = \frac{1}{\exp[(\hbar\omega - \hbar\boldsymbol{k}\cdot\boldsymbol{u}_{\mathrm{D}})/k_{\mathrm{B}}T] - 1}, \tag{2.30}$$

where u_{D} has the dimension of velocity. It is also called the drift velocity of phonon gas. The drift operator in Eq. 2.27 can be written as $D = \partial/\partial t + \boldsymbol{v}_k \cdot \nabla$, where v_{k} is the group velocity of phonons

$$\boldsymbol{v}_k = \frac{\partial \omega}{\partial \boldsymbol{k}} \tag{2.31}$$

In this way, the phonon Boltzmann equation can be approximated by a relaxation form

$$\left(\frac{\partial}{\partial t} + \boldsymbol{v}_k^n \cdot \nabla\right) f^n = \frac{f_0^n - f^n}{\tau_{\mathrm{R}}} + \frac{f_{\mathrm{D}}^n - f^n}{\tau_{\mathrm{N}}} \tag{2.32}$$

In pure crystals at low temperature, the N processes overwhelm R processes. In this case $\tau_{\mathrm{N}} \ll \tau_{\mathrm{R}}$, the phonon distribution function is close to f_{D}. Approximating f with f_{D} can reveal the structure of the solution. In more general cases, the phonon distribution can be expressed as $f = f_{\mathrm{D}} + f_{\mathrm{N}}$, which is further deliberated in

Sect. 2.3.2. The actual relaxation time depends on the phonon frequency, branches, and temperature. For simplicity, the constant values of τ_R and τ_N are used for the present discussion.

Substitution of $f = f_D$ into Eq. 2.32 leads to

$$\left(\frac{\partial}{\partial t} + \boldsymbol{v}_k^n \cdot \nabla\right) f_D^n = \frac{f_0^n - f_D^n}{\tau_R} \tag{2.33}$$

Integral of Eq. 2.33 in the k space with multiplying $\hbar k_i$ or $\hbar\omega v_{ki}$ will give the macroscopic governing equations which predicts the drifting or driftless second sound, respectively [23, 24]. The difference between the two types of second sound can be shown as

$$\frac{v_{\mathrm{II}}'}{v_{\mathrm{II}}} = \frac{\sum\limits_{\text{all }\varepsilon} \langle 0|v^1|\varepsilon\rangle\langle\varepsilon|v^1|0\rangle}{\sum\limits_{\varepsilon=0}^{3} \langle 0|v^1|\varepsilon\rangle\langle\varepsilon|v^1|0\rangle}, \tag{2.34}$$

where $< \alpha|\mathbf{v}|\beta >$ is the matrix element expressed by the eigenvalue, v_{II} and v'_{II} are the velocities of drifting and driftless second sound, respectively. Therefore, the integral path of the driftless second sound, namely multiplying with $\hbar\omega v_{ki}$ covers more eigenvalues than that of the drift second sound. Hardy [23] indicated that "… such a possibility suggests that the different types of second sound should be thought of not as distinct 'modes' of heat propagation, but rather as simply different approximation schemes which lead to the same phenomena."

The second integral method is favorable according to the thermomass theory. In the gas transport theory, multiplying the Boltzmann equation with the molecular momentum, mv, and integrating in the velocity space will lead to the momentum balance equation of fluid. According to the mass energy equivalence, $\hbar\omega$ is the phonon energy and $\hbar\omega/c^2$ is the equivalent mass of phonons. $\hbar\omega v_{ki}/c^2$ is the phonon momentum accompanied by the heat transport. It is an actual momentum rather than the quasi-momentum of phonons, $\hbar k$. Similar to the transport theory of gases, multiplying Eq. 2.33 with $\hbar\omega/c^2$ and $\hbar\omega v_{ki}/c^2$ and making integral, respectively, will give the mass and momentum balance equation of phonon gas. The distinction from the ordinary gas flow is that the phonon gas is also impeded by the R processes, which is reflected by the additional sink term of momentum, namely the collision operator on the right-hand side of Eq. 2.33. It is equivalent to the gas flow in a pipe filled with porous medium rather than empty.

Multiplying Eq. 2.33 with $\hbar\omega/c^2$ or $\hbar\omega v_{ki}/c^2$ and then integrating in the k space yields

$$\frac{\partial \int_{\boldsymbol{k}} f_D^n \hbar\omega^n}{\partial t} + \int\limits_{\boldsymbol{k}} \boldsymbol{v}_k^n \cdot \nabla f_D^n \hbar\omega^n = \frac{\int_{\boldsymbol{k}} (f_0^n - f_D^n)\hbar\omega^n}{\tau_R} \tag{2.35}$$

$$\frac{\partial \int_{\boldsymbol{k}} f_{\mathrm{D}}^{n} \hbar\omega^{n} v_{ki}^{n}}{\partial t} + \int_{\boldsymbol{k}} \boldsymbol{v}_{k}^{n} \cdot \nabla f_{\mathrm{D}}^{n} \hbar\omega^{n} v_{ki}^{n} = \frac{\int_{\boldsymbol{k}} \left(f_{0}^{n} - f_{\mathrm{D}}^{n}\right) \hbar\omega^{n} v_{ki}^{n}}{\tau_{\mathrm{R}}} \tag{2.36}$$

In order to integrate Eqs. 2.35 and 2.36, the feature of f_{D} needs to be discussed. The derivative of f_{D} with respect to the drift velocity is

$$\frac{\partial f_{\mathrm{D}}}{\partial u_{\mathrm{D}j}} = \frac{\hbar k_j}{k_{\mathrm{B}} T} \frac{\exp[(\hbar\omega - \hbar\boldsymbol{k} \cdot \boldsymbol{u}_{\mathrm{D}})/k_{\mathrm{B}} T]}{\{\exp[(\hbar\omega - \hbar\boldsymbol{k} \cdot \boldsymbol{u}_{\mathrm{D}})/k_{\mathrm{B}} T] - 1\}^2}, \tag{2.37}$$

while the derivative of f_0 with respect to frequency is

$$\frac{\partial f_0}{\partial \omega} = \frac{\hbar}{k_{\mathrm{B}} T} \frac{\exp(\hbar\omega/k_{\mathrm{B}} T)}{[\exp(\hbar\omega/k_{\mathrm{B}} T) - 1]^2} \tag{2.38}$$

Therefore, when u_{D} is not large one can approximate that

$$\frac{\partial f_{\mathrm{D}}}{\partial u_{\mathrm{D}j}} \approx k_j \frac{\partial f_0}{\partial \omega} \tag{2.39}$$

In the same manner, the second order derivative of f_{D} with respect to u_{D} is

$$\begin{aligned} \frac{\partial^2 f_{\mathrm{D}}}{\partial u_{\mathrm{D}j}^2} &= \frac{\hbar k_j}{k_{\mathrm{B}} T} \frac{\partial}{\partial u_{\mathrm{D}j}} \frac{\exp[(\hbar\omega - \hbar\boldsymbol{k} \cdot \boldsymbol{u}_{\mathrm{D}})/k_{\mathrm{B}} T]}{\{\exp[(\hbar\omega - \hbar\boldsymbol{k} \cdot \boldsymbol{u}_{\mathrm{D}})/k_{\mathrm{B}} T] - 1\}^2} \\ &= \left(\frac{\hbar k_j}{k_{\mathrm{B}} T}\right)^2 \mathfrak{X} \left[\frac{1}{(\mathfrak{X} - 1)^2} - \frac{2\mathfrak{X}}{(\mathfrak{X} - 1)^3}\right] \\ &\approx k_j^2 \frac{\partial^2 f_0}{\partial \omega^2} \end{aligned}, \tag{2.40}$$

where

$$\mathfrak{X} = \exp[(\hbar\omega - \hbar\boldsymbol{k} \cdot \boldsymbol{u}_{\mathrm{D}})/k_{\mathrm{B}} T] \tag{2.41}$$

With the above results one can make a second order Taylor expansion of f_{D} around f_0

$$\begin{aligned} f_{\mathrm{D}} &= f_0 + \left.\frac{\partial f_{\mathrm{D}}}{\partial \boldsymbol{u}_{\mathrm{D}}}\right|_{\Delta u_D = 0} \Delta \boldsymbol{u}_D + \frac{1}{2} \left.\frac{\partial^2 f_{\mathrm{D}}}{\partial \boldsymbol{u}_{\mathrm{D}}^2}\right|_{\Delta u_D = 0} (\Delta \boldsymbol{u}_{\mathrm{D}})^2 + o\left((\Delta \boldsymbol{u}_{\mathrm{D}})^2\right) \\ &= f_0 + \frac{\partial f_0}{\partial \omega} (\boldsymbol{k} \cdot \boldsymbol{u}_D) + \frac{1}{2} \frac{\partial^2 f_0}{\partial \omega^2} (\boldsymbol{k} \cdot \boldsymbol{u}_{\mathrm{D}})^2 + o\left((\Delta \boldsymbol{u}_{\mathrm{D}})^2\right) \\ &= f_0 + f_+ + f_{++} + o\left((\Delta \boldsymbol{u}_{\mathrm{D}})^2\right) \end{aligned} \tag{2.42}$$

Note that f_0 and f_{++} are both even functions in the k space, while f_+ is an odd function. Substitution of Eq. 2.42 into Eqs. 2.35 and 2.36 gives

$$\frac{\partial \int_k (f_0^n + f_{++}^n)\hbar\omega^n}{\partial t} + \nabla_j \int_k f_+^n \hbar\omega^n v_{kj}^n = -\frac{\int_k f_{++}^n \hbar\omega^n}{\tau_R} \tag{2.43}$$

$$\frac{\partial \int_k f_+^n \hbar\omega^n v_{ki}^n}{\partial t} + \nabla_j \int_k (f_0^n + f_{++}^n)\hbar\omega^n v_{ki}^n v_{kj}^n = -\frac{\int_k f_+^n \hbar\omega^n v_{ki}^n}{\tau_R} \tag{2.44}$$

The second term on the left-hand side of Eq. 2.43 can be detailed as

$$\int_k f_+^n \hbar\omega^n v_{kj}^n = \frac{4}{3} u_{Dj} \int_k f_0^n \hbar\omega^n = \frac{4}{3} u_{Dj} e = q_j \tag{2.45}$$

The second order term, f_{++}, should be much smaller than f_0, so its contribution to the internal energy is temporally neglected here. Further discussion is made in Sect. 4.4.

Note that u_D has a dimension of velocity, which relates to the drift velocity of thermomass as

$$u_D = \frac{3}{4} u_h \tag{2.46}$$

This velocity is called the average drift velocity of phonon gas [25]. However, without defining the mass of phonon gas, the physics of this velocity is ambiguous. Similarly, the velocity

$$u'_D = \frac{3q}{\rho C_V T} = 3u_h \tag{2.47}$$

is defined as the "fluid velocity" of phonon gas by Guyer and Krumahansl [20, 21]. It is also proportional to the drift velocity of thermomass. The difference between u'$_D$ and u_h comes from the average of the homogeneous vector space.

Since the energy conservation is ensured during phonon scatterings, $\hbar\omega$ is the eigenvector of the zero space of the scattering operator, as well as the integral invariance of the Boltzmann equation. Therefore, Eq. 2.43 transforms to

$$\frac{\partial e}{\partial t} + \nabla_j q_j = 0 \tag{2.48}$$

which yields the traditional energy conservation equation. Integral of the second term on the left-hand side of Eq. 2.44 can be divided into an equilibrium and a nonequilibrium part

$$\int_{k} (f_0^n + f_{++}^n)\hbar\omega^n v_{ki}^n v_{kj}^n = \delta_{ij} \int_{k} f_0^n \hbar\omega^n v_{ki}^n v_{kj}^n + \int_{k} f_{++}^n \hbar\omega^n v_{ki}^n v_{kj}^n \tag{2.49}$$

Integration by parts of the second term on the right-hand side of Eq. 2.49 gives

$$\int_{k} f_{++}^n \hbar\omega^n v_{ki}^n v_{kj}^n = \frac{5}{3} u_{\mathrm{D}i} u_{\mathrm{D}j} e \tag{2.50}$$

Inserting Eqs. 2.45, 2.49 and 2.50 into Eq. 2.44 leads to

$$\frac{\partial q_i}{\partial t} + \frac{15}{16}\nabla_j \frac{q_i q_j}{e} + \frac{1}{3}\nabla_j \int_{k} f_0^n \hbar\omega^n \left(v_k^n\right)^2 = -\frac{q_i}{\tau_{\mathrm{R}}} \tag{2.51}$$

where the cubic symmetry is assumed to obtain the third term on the left-hand side of Eq. 2.51.

Equation 2.51 is actually the momentum balance equation of phonon gas. Compared with Eq. 2.20, which can be reformed as

$$\frac{\partial q_i}{\partial t} + \nabla_j \frac{q_i q_j}{e} + \nabla_i p_{\mathrm{h}} = -\chi \frac{q_i}{e} \tag{2.52}$$

one observes that the homogeneous phonon gas pressure and be expressed by the integration of the microscopic phonon properties

$$p_{\mathrm{h}} = \frac{1}{3}\int_{k} f_0^n \frac{\hbar\omega^n}{c^2}\left(v_k^n\right)^2 = \iiint_{\pm\pi/a} f_0^n(\boldsymbol{x}, t, \boldsymbol{k}) \frac{\hbar\omega^n}{c^2}\left(v_{kx}^n\right)^2 \mathrm{d}k_x \mathrm{d}k_y \mathrm{d}k_z \tag{2.53}$$

It is notable that the gas pressure in the kinetic theory is expressed by

$$p_x = \iiint_{\pm\infty} f(\boldsymbol{x}, t, \boldsymbol{v}) m v_x^2 \mathrm{d}v_x \mathrm{d}v_y \mathrm{d}v_z, \tag{2.54}$$

where p is the thermodynamic pressure, f is the local distribution function, m is the molecular mass, and v is the velocity of each molecule. The integral structure of Eq. 2.54 is similar to that of Eq. 2.53. Thus the temperature gradient driving the heat flux has the similar physical meaning to the pressure gradient driving the gas flow.

The thermomass pressure is macroscopically derived from the Debye state equation of dielectric solids. According to the concept of thermomass, the phonons have the real momentum, $\hbar\omega v_{ki}/c^2$. In equilibrium state, the phonons are confined in the medium. Assume that the phonons are reflected at the boundary, i.e., the wave vector is reversed, the pressure of phonons can be calculated in analogy with the ideal gas model. The impulse on the boundary is obtained through the momentum change of phonons. The characteristic velocity of phonons is defined as

$$v_{\mathrm{g}}^{2}=\frac{\sum_{n}\frac{1}{3}\int_{\boldsymbol{k}}f_{0}^{n}\frac{\hbar\omega^{n}}{c^{2}}\left(v_{k}^{n}\right)^{2}}{\sum_{n}\int_{\boldsymbol{k}}f_{0}^{n}\frac{\hbar\omega^{n}}{c^{2}}} \tag{2.55}$$

The denominator of Eq. 2.55 is the density of the equivalent mass of phonon gas, ρ_{h}. The relation between the phonon gas pressure and phonon gas density is

$$p_{\mathrm{h}}=\frac{v_{\mathrm{g}}^{2}}{c^{2}}\rho C_{V}T=v_{\mathrm{g}}^{2}\rho_{\mathrm{h}} \tag{2.56}$$

Comparison with phonon gas pressure given by the thermomass theory yields

$$v_{\mathrm{g}}=\sqrt{\gamma_{\mathrm{G}}C_{V}T} \tag{2.57}$$

Thus v_{g} is not a constant but depends on the temperature. Also, v_{g} is different from the propagation speed of disturbance of phonon gas, v_{s}, which has the form

$$v_{\mathrm{s}}=\sqrt{\frac{dp_{\mathrm{h}}}{d\rho_{\mathrm{h}}}}=\sqrt{2\gamma_{\mathrm{G}}C_{V}T} \tag{2.58}$$

Equation 2.58 is in analogy with the relation between the density, pressure, and sound speed of ideal gas. The difference between v_{s} and v_{g} arises from the compression ratio of phonon gas (thermomass fluid). The propagation of small disturbance in ideal gas is well approximated by the adiabatic thermodynamic process. The compression ratio for the sound speed is determined by the density-pressure relation in an adiabatic process. The density and pressure of phonon gas satisfy

$$p_{\mathrm{h}}\rho_{\mathrm{h}}^{-2}=\mathrm{const} \tag{2.59}$$

Thus the defaulted compression ratio of phonon gas is 2. Since the thermomass density directly depends on the pressure, there is no difference between the isothermal and adiabatic processes. In a non-dispersive medium, assuming that there is only one longitudinal phonon branch (L) and two translational branches (T), with the sound speed v_{L} and v_{T}, respectively, one has [23, 24]

$$v_{\mathrm{g}}^{2}=\frac{1}{3}\frac{v_{\mathrm{L}}^{-1}+2v_{\mathrm{T}}^{-1}}{v_{\mathrm{L}}^{-3}+2v_{\mathrm{T}}^{-3}} \tag{2.60}$$

Thus v_{g} is a weighted average of the sound speed of each phonon branch. If there is no coupling among phonon branches, the disturbance in each branch will propagate separately, causing a rapid dispersion of temperature wave. In this case there is no steady "temperature wave," in other words, the heat wave will not form in such medium. On the other hand, if the phonon can change branches frequently by scattering, namely the phonon in the L branch can transform into phonons in T branches and vice versa, a steady wave propagating with an average sound speed could form in medium. The former situation is close to the case in pure crystals at low temperature. The dispersion of temperature wave is also observable in MD simulations. At higher temperatures, the scattering among high frequency phonons rapidly redistributes the energy and momentum of phonon branches. In this case, the collective behavior of phonons is more likely a gas, which is describable by a continuous model. Equation 2.60 is actually the speed of the driftless second sound proposed by Hardy [23, 24], v'_{II}. For the drift second sound, the speed is

$$v_{II} = \frac{1}{3}\frac{v_{\mathrm{L}}^{-3} + 2v_{\mathrm{T}}^{-3}}{v_{\mathrm{L}}^{-5} + 2v_{\mathrm{T}}^{-5}} \tag{2.61}$$

The difference between v_{II} and v'_{II} originates from the multipliers used in integration, which are $\hbar k$ and $\hbar\omega v_{ki}/c^2$, respectively. It induces different weight for the sound speeds of phonon branches. Enz [26] pointed out that distinction of the driftless second sound from the drift one is decided by the fact of whether the core variable during second sound propagation is the quasi-momentum or energy flux (heat flux). It can be inferred that the drift second sound is more likely to happen in the pure crystal at low temperature. However, the dispersion is strong for heat waves. It can only maintain a stable wave within a short propagation distance. For medium at higher temperatures, the driftless second sound is possible. In this case, dissipation during heat wave propagation is strong. The heat wave is easily merged in the noise caused by R scatterings. If a medium can satisfy the conditions of large MFP for R processes and small N processes, the differences among sound speeds of different phonon branches are small or the propagation is dominated by one phonon branch while other branches contribute little to the heat transfer, the heat wave is more likely to be observable. It is expected that low-dimensional material (such as CNT and graphene) could be good candidates for measuring the heat wave.

Equation 2.51 indicates that the phonon gas pressure plays the role of driving force of heat conduction. For a specific material, the phonon gas pressure can be obtained either from the macroscopic approach, i.e., Eq. 2.17, or the microscopic expression, Eq. 2.53. The phonon gas pressure is difficult to be directly measured. The characteristic relaxation time is the general variable detected in experiments and simulations. Based on Eq. 2.51 one can derive the relation between the relaxation time and the phonon gas pressure

$$\tau_{\mathrm{R}} = \frac{\kappa}{c^2 \mathrm{d}p_{\mathrm{h}}/\mathrm{d}T} \tag{2.62}$$

For the silicon at 300 K, with the properties chosen as κ = 149 W m^{-1} K^{-1}, C_{v} = 704.6 J kg^{-1} K^{-1}, ρ = 2330 kg m^{-3}, γ = 1.5, the first algorithm gives a relaxation time of 1.4 × 10^{-10}s, while the second algorithm yields 0.5 × 10^{-10}s.

The experiments indicate that the relaxation time is 1.5×10^{-10}s [27]. Thus the results obtained through various methods are of the same magnitude.

The three terms on the left-hand side of Eq. 2.51 come from f_+, f_{++} and f_0 respectively. If one only retains the third term, then Eq. 2.51 reduces to the Fourier's heat conduction law. If the terms from f_0 and f_+ are reserved, Eq. 2.51 transforms into the CV model describing the transient heat wave propagation. If all the three terms are fully considered, namely taking the second order expansion of the displaced Planck distribution, one obtains a governing equation similar to the momentum balance equation of phonon gas (Eq. 2.20). In ordinary cases, f_{++} is much smaller than f_0 and is thereby negligible. In extreme conditions such as high heat flux density, the contribution of f_{++} should be considered.

The second term on the left-hand side of Eq. 2.51 (convection term) has a coefficient 15/16. The corresponding term in the momentum balance equation of phonon gas should be unity (cf. Eq. 2.52). This distinction can be analyzed from the phonon energy variation caused by the Doppler Effect. Note that the coefficient 4/3 in Eq. 2.45 comes from the integration by parts

$$\begin{aligned}
\int_{\boldsymbol{k}} f_+^n \hbar\omega^n v_{kj} &= u_{\mathrm{D}j} \int_{\boldsymbol{k}} f_0^n \hbar\omega^n + \int_{\boldsymbol{k}} \hbar(\boldsymbol{k} \cdot \boldsymbol{u}_{\mathrm{D}}) \frac{\partial \omega^n}{\partial k_j} f_0^n \\
&= u_{\mathrm{D}j} \int_{\boldsymbol{k}} f_0^n \hbar\omega^n + \frac{1}{3} u_{\mathrm{D}j} \int_{\boldsymbol{k}} \hbar\omega^n f_0^n \\
&= u_{\mathrm{D}j} \int_{\boldsymbol{k}} f_0^n \hbar\omega^n + B u_{\mathrm{D}j} \int_{\boldsymbol{k}} \hbar\omega^n f_0^n
\end{aligned} \tag{2.63}$$

This integration assumes the cubic symmetry and agrees with Sussmann's results [20]. Equation 2.63 contains two parts. One is a uniform drift of the equilibrium part f_0 with a velocity u_{Dj}. The second part is induced by the derivative of the phonon energy, $\hbar\omega$, with respect to the wave vector k, which is measured by the coefficient B. The coefficient 15/16 in Eq. 2.51 is $(1 + 2B)/(1 + B)^2$, which is always less than unity. Therefore, this additional coefficient rises from the Doppler Effect during the drift motion of phonon gas. From this perspective, the phonon gas is slightly different from the real gas. The phonon energy varies due to dispersion causing the "eclipse" of the convection term. In a non-dispersive medium, the frequency is independent of k. In this case $B = 0$, and Eq. 2.51 will have exactly the same form as Eq. 2.52.

2.3.2 Chapman–Enskog Expansion

In Sect. 2.3.1 it is assumed that the phonon distribution function is approximated by f_{D}, which is reasonable only in pure crystals at low temperature. In other cases the

Umklapp scattering, impurity scattering, and other momentum breaking processes will continuously draw f away from f_D, which relaxes back to f_D with a relaxation time τ_N.

The second derivative of heat flux is proportional to τ_N in Eq. 1.7 with a scale coefficient $\tau_R\tau_N{v_s}^2/5$. When τ_R is much larger than τ_N, Eq. 2.32 can be written as

$$f_D^n - f^n = \tau_N(\partial/\partial t + \boldsymbol{v}_k \cdot \nabla)f^n \tag{2.64}$$

This form is close to Sussmann's assumption [22]. Since f_D can be regarded to represent the uniform drift motion of phonon gas, the second derivative term caused by τ_N thereby comes from the nonuniform motion of phonon gas. In this manner, the τ_N term in the Boltzmann equation gives the additional friction force induced by the local gradient of drift velocity. If one introduces a heterogeneous term into the thermomass pressure (Eq. 2.17) or the phonon gas pressure (Eq. 2.53) in analogy with the shear stress elements, the second derivative of heat flux then emerges as the viscous dissipation term containing the Laplacian of velocity in Navier–Stokes equations.

For more general cases, Banach and Jiaung et al. [28, 29] solved the phonon Boltzmann equation with the Chapman–Enskog expansion. The common points of these methods are: (1) The phonon distribution function is expressed by an expansion around f_D

$$f = f_D + Knf_1 + Kn^2 f_2 + \cdots \tag{2.65}$$

where Kn is the Knudsen number, which is the ratio of the average mean free path of particles over the characteristic size of flow region. (2) The first order Chapman–Enskog expansion f_1 is proportional to τ_N. (3) The first order Chapman–Enskog expansion will lead the second order derivative of heat flux in the governing equation. This is similar to the basic assumption in the derivation of Navier–Stokes equations.

Sussmann and Thellung [22] assumed that the phonon distribution function has the form

$$f = f_D - \tau_N(\partial/\partial t + \boldsymbol{v}_k \cdot \nabla)f_D \tag{2.66}$$

It also satisfies the three essential features of Chapman–Enskog expansion.

Without the Chapman–Enskog expansion, Banach et al. [28] obtained the governing equation of heat conduction in three-dimensional space as

$$\frac{\partial q_i}{\partial t} + \frac{1}{3}\nabla_j \int\limits_{\boldsymbol{k}} f_0^n \hbar\omega \left(v_k^n\right)^2 + \nabla_j M^{ij} = -\frac{q_i}{\tau_R} \tag{2.67}$$

where the zeroth order approximation of M_{ij} satisfies

$$M_0^{ij} = \frac{3}{2e + \sqrt{4e^2 - 3|\boldsymbol{q}|^2 / (v_k^n)^2}} \left(q_i q_j - \frac{1}{3} \delta_{ij} \right) \quad (2-68)$$

The drift velocity of phonon gas, u_h, is usually small. Therefore, $(q/v_s)^2$ is negligible compared with $4e^2$. Compared with Eq. 2.49, Eq. 2.67 can transform to

$$\frac{\partial q_i}{\partial t} + \frac{3}{4} \nabla_j \frac{q_i q_j}{e} + \frac{1}{3} \nabla_j \int_{\boldsymbol{k}} f_0^s \hbar\omega (v_k^n)^2 = -\frac{q_i}{\tau_R} \quad (2.69)$$

Equation 2.69 has the same structure as Eq. 2.51, with the only difference in the coefficient of the convection term. This distinction is caused by overestimation of the Doppler effect during integration.

Therefore, one can conclude that the second order derivative term rises from the Chapman–Enskog expansion around f_D. The magnitude of Chapman–Enskog expansion is proportional to Kn. So this effect should be considered in the condition of large Kn, for e.g., the heat conduction in nanosystems.

2.3.3 *Eigenvalue Analysis*

The eigenvalue analysis has been used to solve the phonon Boltzmann equation by Guyer and Krumhansl [20] and Hardy [23]. The phonon distribution function is converted into symmetry form

$$f^*(\boldsymbol{k}, x, t) = f(\boldsymbol{k}, x, t) 2 \sin h\left(\frac{1}{2} x_k\right), \quad (2.70)$$

where

$$x_k = \frac{\hbar\omega}{k_B T} \quad (2.71)$$

With this the phonon Boltzmann equation (Eq. 2.27) turns to

$$Df^* = (R^* + N^*) f^* \quad (2.72)$$

It is assumed that the zero subspace of N^* has four and only four eigenvectors

$$\begin{aligned} |\eta_0\rangle &= \mu x_k \left(2\sinh\frac{1}{2}x_k\right)^{-1} \\ |\eta_{1x}\rangle &= \lambda_x q_x \left(2k_{\mathrm{B}}T\sinh\frac{1}{2}x_k\right)^{-1} \\ |\eta_{1y}\rangle &= \lambda_y q_y \left(2k_{\mathrm{B}}T\sinh\frac{1}{2}x_k\right)^{-1} \\ |\eta_{1z}\rangle &= \lambda_z q_z \left(2k_{\mathrm{B}}T\sinh\frac{1}{2}x_k\right)^{-1} \end{aligned} \tag{2.73}$$

These four eigenvectors correspond to the internal energy and the heat flux in three directions in a Cartesian coordinate system. $|\eta_0\rangle$ is the eigenvector of the zero subspace of both N* and R*. $|\eta_{1x}\rangle$, $|\eta_{1y}\rangle$ and $|\eta_{1z}\rangle$ are only the eigenvectors of the zero subspace of N* but do not belong to the zero subspace of R*, because R* is the momentum breaking operator that does not conserve the quasi-momentum. Assuming that all other eigenvectors have nonzero eigenvalues, f^* can be written as the linear combination of the eigenvectors of N^*, which give the solution of the phonon Boltzmann equation

$$\frac{\partial \boldsymbol{q}}{\partial t} + \frac{1}{3}(v_{\mathrm{s}})^2\nabla E = -\frac{\boldsymbol{q}}{\tau_{\mathrm{R}}} + \frac{\tau_{\mathrm{N}}(v_{\mathrm{s}})^2}{5}\left(\nabla^2 + \zeta\nabla\nabla\cdot\right)\boldsymbol{q}, \tag{2.74}$$

where $\zeta = 2$ in [20] while $\zeta = 1/3$ in [23]. The second order derivative comes from the eigenvectors with nonzero eigenvalues. For $\beta = 1, 2, 3$, the expansions of the phonon distribution function around f_{D} is

$$\sum_{\alpha=0}^{3}\sum_{\sigma\geq 4}\sum_{\mu\geq 4}\frac{\langle\beta|\boldsymbol{D}+\boldsymbol{R}|\sigma\rangle\langle\mu|\boldsymbol{D}+\boldsymbol{R}|\alpha\rangle}{N^{\mu}}f_{\alpha}, \tag{2.75}$$

where $\langle\alpha|\boldsymbol{D}+\boldsymbol{R}|\beta\rangle$ is the matrix element of the eigenvalues of $\boldsymbol{D}+\boldsymbol{R}$, N^{μ} denotes the μth eigenvalue of the normal process collision matrix, and $\boldsymbol{R}$ is the resistive collision operator. Since the drift operator D contains the first order derivative of space, Eq. 2.75 introduces the second order derivative term in the governing equation. If the eigenvectors $|\sigma\rangle$ and $|\mu\rangle$ have zero eigenvalues, the second order derivative will vanish simultaneously. Therefore, Eq. 2.75 indicates that existence of the second order derivative in the governing equation requires that all the eigenvectors except $|\eta_0\rangle$, $|\eta_{1x}\rangle$, $|\eta_{1y}\rangle$ and $|\eta_{1z}\rangle$ should have nonzero eigenvalues with respect to $\boldsymbol{D}+\boldsymbol{R}$.

The phonon distribution function is approximated by f_{D} in Sect. 2.3.1. Based on Krumhansl's transformation [30], the first order expansion of f_{D}, namely f_0+f_+, is one of the eigenvectors of the zero subspace of N. In this sense, if $|\eta_{1x}\rangle$, $|\eta_{1y}\rangle$ and $|\eta_{1z}\rangle$ are strictly proportional to the heat flux, they will have small but nonzero eigenvalues.

If their eigenvalue is strictly zero, they should contain the higher order term of heat flux. If the higher order terms of heat flux are considered in $|\eta_{1x}\rangle$, $|\eta_{1y}\rangle$ and $|\eta_{1z}\rangle$, it will introduce the convection term in the governing equation of phonon gas dynamics. This is the reason for the distinction between Eqs. 2.51 and 2.74.

2.4 Conclusion

Based on Einstein's mass energy equivalence, the thermal energy stored in the medium contributes to its rest mass, namely the thermomass. The thermomass is a real mass. It has the inertia and gravity effects. The thermomass moves relatively to the rest lattices in heat conduction. The mass flow rate of thermomass corresponds to the heat flux. Based on the generalized mass flux and momentum flux, it can be derived that the motion of thermomass satisfies Newton's law of motion as long as the conduction medium and the thermomass move much slower than the speed of light.

Phonons are the main heat carriers in dielectric solids. The aggregate of phonons in pure dielectric crystals can be regarded as a phonon gas. The dynamic variables such as the equation of state, density, drift velocity, and momentum of phonon gas can be derived from the thermomass theory. The balance equations of the mass and momentum of phonon gas lead to the energy conservation relation and the general heat conduction law. In bulk materials, the friction force on the phonon gas obeys the Darcy's law, i.e., proportional to the drift velocity of phonon gas. The viscosity of phonon gas causes the boundary to additionally impede the phonon gas in nanosystems. In this case the friction force on the phonon gas contains not only the Darcy term proportional to the drift velocity but also the Brinkman term proportional to the second order derivative of the drift velocity.

The dynamic governing equations of phonon gas can be microscopically derived based on the phonon Boltzmann equation. When the relaxation time of R processes is much larger than that of N processes, the phonon distribution function can be approximated by the displaced Planck distribution, f_D. When only the zeroth order of f_D is considered, the Boltzmann equation will give the traditional Fourier's law. Considering the first order expansion of f_D will give the CV model. If the second order of expansion is reserved, one obtains the momentum balance equation phonon gas containing the convection term, which agrees with the macroscopic derivation based on the thermomass theory (Eq. 2.20). If the relaxation time of R processes is not big enough, then the Chapman–Enskog expansion can be implemented around f_D. In this case the second order derivative of heat flux will be introduced into the heat conduction equation. The eigenvalue analysis also obtains the second order derivative of heat flux. However, it assumes there are only four eigenvectors in the zero subspace of N operator, which resultantly eliminates the convection term in the heat conduction equation.

References

1. Moller C (1972) The theory of relativity. Clarendon Press, Oxford
2. Misner CM (1973) Gravitation. WH Freeman and Company, San Francisco
3. Schroder UE (1990) Special relativity. World Scientific, Singapore
4. Rindler W (1982) Introduction to special relativity. Clarendon Press, Oxford
5. Einstein A, Infield L (1938) The evolution of physics: the growth of ideas from early concepts to relativity and quanta. Simon and Schuster, New York
6. Taylor EF, Wheeler JA (1966) Spacetime physics. W.H. Freeman and Company, New York
7. Kittel C (1996) Introduction to solid state physics, 7th edn. Wiley, New York
8. Reissland JA (1973) The physics of phonons. Wiley, London
9. Guo ZY, Cao BY, Zhu HY et al (2007) State equation of phonon gas and conservation equations for phonon gas motion. Acta Phys Sin 56(6):3306–3312
10. Guo ZY, Cao BY (2008) A general heat conduction law based on the concept of motion of thermal mass. Acta Phys Sin 57(7):4273–4281
11. Cao BY, Guo ZY (2007) Equation of motion of a phonon gas and non-Fourier heat conduction. J Appl Phys 102(5):053503
12. Tzou DY, Guo ZY (2010) Nonlocal behavior in thermal lagging. Int J Therm Sci 49(7):1133–1137
13. Guo ZY, Hou QW (2010) Thermal wave based on the thermomass model. J Heat Transfer 132 (7):072403
14. Wang M, Guo ZY (2010) Understanding of temperature and size dependences of effective thermal conductivity of nanotubes. Phys Lett A 374(42):4312–4315
15. Wang M, Yang N, Guo ZY (2011) Non-Fourier heat conductions in nanomaterials. J Appl Phys 110(6):064310
16. Wang M, Shan X, Yang N (2012) Understanding length dependences of effective thermal conductivity of nanowires. Phys Lett A 376(46):3514–3517
17. Wang HD, Cao BY, Guo ZY (2010) Heat flow choking in carbon nanotubes. Int J Heat Mass Transf 53(9):1796–1800
18. Christov CI, Jordan PM (2005) Heat conduction paradox involving second-sound propagation in moving media. Phys Rev Lett 94:154301
19. Müller I, Ruggeri T (1993) Extended thermodynamics. Springer, New York
20. Guyer RA, Krumhansl JA (1966) Solution of the linearized phonon Boltzmann equation. Phys Rev 148(2):766–778
21. Guyer RA, Krumhansl JA (1966) Thermal conductivity, second sound, and phonon hydrodynamic phenomena in nonmetallic crystals. Phys Rev 148(2):778
22. Sussmann JA, Thellung A (1963) Thermal conductivity of perfect dielectric crystals in the absence of umklapp processes. Proc Phys Soc 81(6):1122
23. Hardy RJ (1970) Phonon Boltzmann equation and second sound in solids. Phys Rev B 2 (4):1193
24. Hardy RJ, Albers DL (1974) Hydrodynamic approximation to the phonon Boltzmann equation. Phys Rev B 10(8):3546
25. Gurevich VL (1986) Transport in phonon systems. North-Holland, Amsterdam
26. Enz CP (1968) One-particle densities, thermal propagation, and second sound in dielectric crystals. Ann Phys 46(1):114–173
27. Sahasrabudhe GG, Lambade SD (1999) Temperature dependence of the collective phonon relaxation time and acoustic damping in Ge and Si. J Phys Chem Solids 60 (6):773–785
28. Banach Z, Larecki W (2008) Chapman-Enskog method for a phonon gas with finite heat flux. J Phys A Math Theor 41(37):375502

29. Jiaung WS, Ho JR (2008) Lattice-Boltzmann modeling of phonon hydrodynamics. Phys Rev E 77(6):066710
30. Krumhansl JA (1965) Thermal conductivity of insulating crystals in the presence of normal processes. Proc Phys Soc 85(5):921

Chapter 3
General Entropy Production Based on Dynamical Analysis

Abstract The classical expression for entropy production is a bilinear product of generalized forces and fluxes. The combination of the Cattaneo-Vernotte model with the classical entropy expression gives a non-quadratic form, which needs to be mended to avoid the paradox of negative entropy production. Based on the thermomass theory, it is shown that the entropy production corresponds to the dissipation of the mechanical energy of thermomass flow. Therefore, the generalized forces in the entropy production should be the friction force rather than the driving force. The friction force is proportional to the heat flux. The general entropy production is thus derived as a quadratic form of heat flux, avoiding the paradox of the negative entropy production. The generalized forces and fluxes in other irreversible transport processes are investigated following the similar framework. The friction forces, driving forces and drift velocities are clarified for these transports and the general entropy production for various transport processes are derived.

3.1 Extended Entropy Production

The entropy production is the key quantity in the irreversible thermodynamics. It is the cornerstone for the derivation of Onsager reciprocal relation and the minimum entropy production principle. In Sect. 1.3 it has been pointed out that the traditional expression in irreversible thermodynamics will face the negative entropy production paradox in case of the transient non-Fourier heat conduction. The extended irreversible thermodynamics (EIT) introduces new state variables to modify the entropy production expression. The modified (extended) entropy production keeps semi-positive definite in non-Fourier heat conductions, which avoids the paradox of breaking second law of thermodynamics.

Consider a non-deformable solid (constant volume system). If q is upgraded to an independent state variable, the entropy density, s, in unit volume is [1–16]

Y. Dong, *Dynamical Analysis of Non-Fourier Heat Conduction and Its Application in Nanosystems*, Springer Theses,
DOI 10.1007/978-3-662-48485-2_3

$$\mathrm{d}s = \left(\frac{\partial s}{\partial e}\right)\mathrm{d}e + \left(\frac{\partial s}{\partial \boldsymbol{q}}\right) \cdot \mathrm{d}\boldsymbol{q} \tag{3.1}$$

where e is the density of internal energy. EIT assumes that Eq. 3.1 can be linearized as

$$\mathrm{d}s = \theta^{-1}\mathrm{d}e - T^{-1}\rho^{-1}\alpha_{10}\boldsymbol{q} \cdot \mathrm{d}\boldsymbol{q} \tag{3.2}$$

where θ is the non-equilibrium temperature

$$\theta^{-1}(e, \boldsymbol{q}) = \left(\frac{\partial s}{\partial e}\right)_{\boldsymbol{q}} \tag{3.3}$$

Remind that the internal energy, entropy and temperature have the relation in traditional thermodynamics

$$T^{-1} = \frac{\partial s}{\partial e} \tag{3.4}$$

The derivative of entropy in Eq. 3.2 with respect to internal energy is no longer the inverse of T. Instead it's the inverse of θ. Here the temperature and entropy are both extended in non-equilibrium systems. They are not only the function of the internal energy density, e, but also the function of the heat flux, q. The second term on the right hand side of Eq. 3.2 has a minus sign. If the coefficient α_{10} is positive, Eq. 3.2 indicates that the heat flux in system would reduce the entropy.

The time derivative of Eq. 3.2 is

$$\frac{\mathrm{d}s}{\mathrm{d}t} = -\theta^{-1}\nabla \cdot \boldsymbol{q} - T^{-1}\alpha_{10}\boldsymbol{q} \cdot \frac{\mathrm{d}\boldsymbol{q}}{\mathrm{d}t} \tag{3.5}$$

If the entropy flux J^s is defined in a traditional form

$$\boldsymbol{J}^s = \theta^{-1}\boldsymbol{q} \tag{3.6}$$

Substituting it into Eq. 3.5 yields the local entropy production rate

$$\sigma^s = \boldsymbol{q} \cdot \left(\nabla\theta^{-1} - T^{-1}\alpha_{10}\dot{\boldsymbol{q}}\right) \tag{3.7}$$

In the classical irreversible thermodynamics, the entropy production is generally the bilinear product of the generalized forces and fluxes. The heat flux in Eq. 3.7 can be regarded as the generalized flux with other terms the generalized force. Assume that the generalized force is linear to the flux, one has

$$\boldsymbol{q} = K\left(\nabla\theta^{-1} - T^{-1}\alpha_{10}\dot{\boldsymbol{q}}\right) \tag{3.8}$$

In steady state, Eq. 3.8 should recover the Fourier's heat conduction law. Thus the coefficient of Eq. 3.8 can be determined as $K = \kappa\theta^2$. On the other hand, if Eq. 3.8 recovers the Cattaneo-Vernotte (CV) model in the fast transient heat conduction, one has

$$\alpha_{10} = \frac{\tau}{\kappa T} \tag{3.9}$$

Substitution of Eq. 3.9 into Eq. 3.7 gives

$$\sigma^s = \boldsymbol{q} \cdot \left(\frac{\nabla T}{T^2} - \frac{\tau}{\kappa T^2}\frac{\partial \boldsymbol{q}}{\partial t}\right) = \frac{q^2}{\kappa T^2} \tag{3.10}$$

Note that the second equation in Eq. 3.10 also adopts the CV model. Inserting Eq. 3.9 into Eq. 3.5 further leads to the extended entropy which is compatible with the CV model

$$\mathrm{d}s = T^{-1}\mathrm{d}e - \frac{\tau}{\kappa T^2}\boldsymbol{q} \cdot \mathrm{d}\boldsymbol{q} + o\left(q^2\right) \tag{3.11}$$

$$s = \rho C_\mathrm{V} \ln T - \frac{1}{2}\frac{\tau}{\kappa T^2}\boldsymbol{q} \cdot \boldsymbol{q} \tag{3.12}$$

Here the terms on the higher order of q^2 are neglected. Thus the non-equilibrium temperature θ is close to the equilibrium one, T.

The entropy production derived in EIT (Eq. 3.10) is proportional to the square of heat flux. It keeps semi-positive definite in non-Fourier heat conduction. In the condition of Fourier heat conduction, $q = -\kappa\nabla T$. Then Eq. 3.10 is equivalent to the classical expression in irreversible thermodynamics. However, the generalized force in Eq. 3.7 is derived based on the presupposition of CV model. The generality of the extended expression for entropy production is uncertain when other types of non-Fourier heat conduction laws are used. Therefore, it is desired to explore the physics of entropy production in irreversible processes from the first principle and thereby develop the general expression of entropy production.

3.2 Heat Conduction

According to the thermomass theory, the picture of heat conduction is a porous flow. The potential and kinetic energies, e_p and e_k, are

$$e_{\mathrm{p}} = p_{\mathrm{h}} = \frac{\gamma_{\mathrm{G}}\rho}{c^2}(C_{\mathrm{V}}T)^2 \tag{3.13}$$

$$e_{\mathrm{k}} = \frac{1}{2}\rho_{\mathrm{h}}u_{\mathrm{h}}^2 = \frac{1}{2}\frac{q^2}{\rho C_{\mathrm{V}}Tc^2} \tag{3.14}$$

In an equilibrium system, the thermomass does not move. The kinetic energy is zero. The potential energy relates to the internal energy as

$$\mathrm{d}e = \frac{1}{\xi}\mathrm{d}e_{\mathrm{p}} = \rho C_{\mathrm{V}}\mathrm{d}T \tag{3.15}$$

where the dimensionless coefficient ξ is

$$\xi = \frac{2\gamma_{\mathrm{G}}C_{\mathrm{V}}T}{c^2} = \frac{v_{\mathrm{s}}^2}{c^2} \tag{3.16}$$

It is the square of the ratio of the sound speed over the light speed.

As the heat conduction occurs, the thermomass flow possesses a kinetic energy. The total mechanical energy, e_{h}, is the sum of the potential and kinetic energies

$$e_{\mathrm{h}} = e_{\mathrm{p}} + e_{\mathrm{k}} \tag{3.17}$$

Since the drift velocity of thermomass, u_{h}, is usually small, the potential energy is much larger than the kinetic energy. In extreme cases, such as the heat wave propagation and the ultra-high heat flux, the kinetic energy of thermomass could be non-negligible. In these cases the Fourier's law fails. The heat flux, q, becomes independent of the temperature profile in the system.

Note that the entropy production is the measurement of the irreversibility during transport processes. From a mechanical perspective, the irreversibility of fluid flow comes from the dissipation of mechanical energy causfed by the friction. Since the friction on the thermomass flow is proportional to its drift velocity (the present discussion is limited in the bulk material), the local dissipation rate of mechanical energy of thermomass is the product of friction and drift velocity

$$\frac{\mathrm{d}e_{\mathrm{h}}}{\mathrm{d}t} + \nabla\cdot\boldsymbol{J}_{\mathrm{h}} = \boldsymbol{f}_{\mathrm{h}}\cdot\boldsymbol{u}_{\mathrm{h}} = -\frac{2\gamma_{\mathrm{G}}C_{\mathrm{V}}q^2}{\kappa c^2} \tag{3.18}$$

where J_{h} is the flux of mechanical energy passing through a cross section. Equation 3.18 represents the dissipation rate of mechanical energy of thermomass in a unit volume. The entropy production rate is

$$\sigma^s = -\frac{\boldsymbol{f}_{\mathrm{h}}\cdot\boldsymbol{u}_{\mathrm{h}}}{T\xi} = -\frac{1}{T}\boldsymbol{f}_{\mathrm{h}}\cdot\boldsymbol{u}_{\mathrm{h}}\bigg/\left(\frac{2\gamma_{\mathrm{G}}C_{\mathrm{V}}T}{c^2}\right) = \frac{1}{\kappa T^2}\boldsymbol{q}\cdot\boldsymbol{q} \tag{3.19}$$

The coefficient ξ is the ratio of thermomass mechanical energy over the internal energy. It should enter into the denominator because the entropy measures the disability of the conversion from the internal energy to other forms of energies. The macroscopic friction force on the heat conduction, F_h, is

$$\boldsymbol{F}_{\mathrm{h}} = -\frac{\rho C_{\mathrm{V}}\boldsymbol{q}}{\kappa} = \boldsymbol{f}_{\mathrm{h}} \Big/ \left(\frac{2\gamma_{\mathrm{G}}C_{\mathrm{V}}T}{c^2}\right) = \boldsymbol{f}_{\mathrm{h}}/\xi \tag{3.20}$$

The entropy production can be written as

$$\sigma^s = -\frac{1}{T}\boldsymbol{F}_{\mathrm{h}} \cdot \boldsymbol{u}_{\mathrm{h}} = \frac{1}{\kappa T^2}\boldsymbol{q} \cdot \boldsymbol{q} \tag{3.21}$$

Equation 3.21 has the same form as the EIT expression, Eq. 3.10. In the non-Fourier heat conduction, it keeps semi-positive definite. The benefit of the present dynamic analysis is that the only assumption in derivation is the linearity between the friction force and drift velocity (thereby heat flux). It doesn't specify which type of heat conduction model is used.

The generalized force proposed by EIT is [3]

$$\boldsymbol{X} = \nabla\theta^{-1} - \frac{\tau}{\kappa\theta^2}\frac{\partial \boldsymbol{q}}{\partial t} \tag{3.22}$$

Substitution of the relaxation time of thermomass, τ_{TM}, in Eq. 3.22 gives

$$\begin{aligned} -\boldsymbol{X} &= \frac{\nabla T}{T^2} + \frac{1}{2\gamma_{\mathrm{G}}\rho C^2 T^3}\frac{\partial q}{\partial t} \\ &= \frac{c^2}{2\gamma_{\mathrm{G}}\rho C^2 T^3}\left[\nabla p_{\mathrm{h}} + \frac{\partial(\rho_{\mathrm{h}}\boldsymbol{u}_{\mathrm{h}})}{\partial t}\right] = \frac{1}{\rho C_{\mathrm{V}} T^2 \xi}\boldsymbol{f}_{\mathrm{h}} \end{aligned} \tag{3.23}$$

Therefore it can be seen that the generalized force in EIT corresponds to the friction force in the thermomass theory. In particular, the first term in Eq. 3.22 is the temperature gradient, which relates to the driving force in thermomass theory. The second term in Eq. 3.22 is equivalent to the inertia force of thermomass. If the time partial derivative in Eq. 3.22 is replaced by the material derivative D/Dt, then the CV model will have th same form as the momentum balance equation of thermomass (Eq. 2.10)

$$\tau_{\mathrm{TM}}\frac{\mathrm{D}\boldsymbol{q}}{\mathrm{D}t} + \kappa\nabla T + \boldsymbol{q} = 0 \tag{3.24}$$

In the classical irreversible thermodynamics, the generalized forces are defined as the driving forces, since they usually have the form of the gradient of thermodynamic potentials. They are the cause of the generalized fluxes. From the thermomass perspective, the generalized forces in the entropy production should be the friction forces, with a true unit of volume force, N/m^3. Thus the inapplicability of the classical expression of entropy production rises from the misuse of the generalized forces. It is the friction force rather than the driving force that causes dissipation and irreversibility. Based on the thermomass theory, as long as the friction force is in the opposite direction of the driving force, the entropy production is positive definite. EIT obtains the expression of the general entropy production, wherein the definition of the generalized forces and fluxes are still ambiguous. The thermomass theory explicitly defines the forces and fluxes in the heat conduction process. The friction force on thermomass is a real force with the unit of N/m^3. As the entropy production is modified, the expression of entropy could be influenced as well, which will be discussed in detail in Sect. 4.3.

3.3 Mass Diffusion

Without the chemical reaction, the mass conservation equation in the mass diffusion process is [17]

$$\rho \frac{\partial c_k}{\partial t} = -\nabla \cdot \boldsymbol{J}_k \tag{3.25}$$

Here the substance is assumed to be in a closed tank with no barycentric motion. The Fick's law indicates that the mass diffusion flux of a component, $\boldsymbol{J}_\mathrm{k}$, is proportional to the mass ratio, c_k (cf. Eq. 1.16b). In the binary solutions or dilute systems, D_kj can be lumped into D_k for simplicity [18]. In the classical theory, the entropy production in a N component dilute solution due to mass diffusion is

$$\sigma^s = -\sum_{k=1}^{N} \boldsymbol{J}_k \cdot \nabla\left(\frac{\mu_k}{T}\right) \tag{3.26}$$

The chemical potential, μ_k, can be written as the function of mass ratio [19]

$$\mu_k = \frac{RT}{M_k} \ln \alpha_k c_k + \mathrm{const} \tag{3.27}$$

where R is the ideal gas constant, M is the relative molecular mass, α_k is the activity coefficient. For the ideal solution such as isomer/isotopic mixtures and mass diluted systems, α_k will be close to unity. In this case the entropy production can be written as

$$\sigma^s = -\sum_k \boldsymbol{J}_k \cdot \frac{R}{M_k c_k} \nabla c_k \tag{3.28}$$

This expression is positive definite as long as the Fick's law holds.

Without the barycentric motion, the flux of the component diffusion can be reformed with the term of a diffusion velocity

$$\boldsymbol{J}_k = \rho_k \boldsymbol{u}_k \tag{3.29}$$

Inserting Eq. 3.29 into Eq. 3.28 yields

$$\sigma^s = -\sum_k \rho_k \boldsymbol{u}_k \cdot \frac{R}{M_k c_k} \nabla c_k = -\sum_k \boldsymbol{u}_k \cdot \frac{R}{M_k} \nabla \rho_k \tag{3.30}$$

In a isothermal system, it can be further evolved as

$$\sigma^s = -\sum_k \frac{1}{T} \boldsymbol{u}_k \cdot \nabla \frac{\rho_k RT}{M_k} = -\frac{1}{T} \sum_k \boldsymbol{u}_k \cdot \nabla p_k \tag{3.31}$$

where p_k is the partial pressure of the ideal gas or vapor pressure of the component in solution. In this manner, the classical entropy production for mass diffusion can be defined as the product of the driving forces (gradient of p_k) and the diffusive velocities of components, divided by T. The driving forces are identical to external forces acting on the system. A similar expression of the driving force is adopted by Gallavotti [20] in deriving the microscopic entropy production of a deterministic dynamical system.

The friction force in the mass diffusion process can be determined in a similar way to the heat conduction problem. When the Fick's law holds, the friction force should be balanced with the driving force, namely

$$\boldsymbol{f}_k = \nabla p_k = \frac{\rho RT}{M_k} \nabla c_k \tag{3.32}$$

Hence the friction force is

$$\boldsymbol{f}_k = -\rho_k \frac{RT}{D_k M_k} \boldsymbol{u}_k = -\frac{p_k}{D_k} \boldsymbol{u}_k \tag{3.33}$$

Note that the friction force in the thermomass theory can be written as

$$\boldsymbol{f}_{\mathrm{h}} = -\frac{2\gamma_{\mathrm{G}} \rho^2 C_{\mathrm{V}}^3 T^2}{\kappa c^2} \boldsymbol{u}_{\mathrm{h}} = -2p_{\mathrm{h}} \frac{\rho C_{\mathrm{V}}}{\kappa} \boldsymbol{u}_{\mathrm{h}} \tag{3.34}$$

Equation 3.33 has a similar form to Eq. 3.34. The diffusivity, D_k, has a unit of m^2/s, which is the same as the thermal diffusivity.

Combined with the mass conservation equation, Eq. 3.25, the Fick's law gives the diffusive type equation of the mass concentration, which is in analogy to Eq. 1.3

$$\frac{\partial c_k}{\partial t} = -D_k \nabla^2 c_k \tag{3.35}$$

which indicates an infinitive propagation speed of the disturbance of the mass concentration. To remove the paradox, a relaxation model can be introduced as the CV model

$$\tau_k \frac{\partial \boldsymbol{J}_k}{\partial t} + \boldsymbol{J}_k = -\rho D_k \nabla c_k \tag{3.36}$$

where τ_k is the relaxation time. Inserting Eq. 3.36 into the traditional entropy production, Eq. 3.28, also induces the negative entropy production. Therefore, the classical expression should be modified to avoid this paradox. In heat conduction process, the negative entropy production is caused by inappropriately defining the generalized force as the driving force. If the generalized force in the mass diffusion process is replaced by the friction force, one obtains the general entropy production. For the ideal mixture solution, the general entropy production is

$$\sigma^s = -\frac{1}{T}\sum_{k=1}^{N} \boldsymbol{u}_k \cdot \boldsymbol{f}_k = \frac{1}{T}\sum_{k=1}^{N} \frac{p_k}{D_k} \boldsymbol{u}_k \cdot \boldsymbol{u}_k \tag{3.37}$$

This expression has a quadratic form which ensures the semi-positive definite.

In more general cases, the cross effect of various diffusion flux should be considered. The driving force in Eq. 3.31 cannot be expressed by the partial pressure or vapor pressure of a single component, while the friction in Eq. 3.33 will contain the contribution of multiple diffusion flux. Therefore, Eq. 3.37 can be generalized as

$$\sigma^s = -\frac{1}{T}\sum_{k=1}^{N} \boldsymbol{u}_k \cdot \boldsymbol{f}_k = \frac{1}{T}\sum_{k=1}^{N} \left(\boldsymbol{u}_k \cdot \sum_{j=1}^{N} \Lambda_{kj} \boldsymbol{u}_j \right) \tag{3.38}$$

The coefficient matrix Λ should be symmetry according to the Onsager reciprocal relation.

3.4 Electrical Conduction

For electrical conduction, the entropy production in the classical irreversible thermodynamics is expressed as the production of the electrical driving force (gradient of the electrical potential, φ_{e}) and the resultant flux (electric current, i)

$$\sigma^s = \frac{1}{T} \boldsymbol{i} \cdot \nabla \varphi_{\mathrm{e}} \tag{3.39}$$

The electrical current can be rewritten as a velocity form

$$\boldsymbol{i} = \rho_{\mathrm{e}} \boldsymbol{u}_{\mathrm{e}} \tag{3.40}$$

where ρ_{e} is the density of charge and $\boldsymbol{u}_{\mathrm{e}}$ is the drift velocity of charge carriers. In ordinary cases the electrical conduction obeys the Ohm's law. It is actually the case of the balance between driving and friction force. Therefore one has

$$\boldsymbol{f}_{\mathrm{e}} = -\rho_{\mathrm{e}}(-\nabla \varphi_{\mathrm{e}}) = -\boldsymbol{i} r_{\mathrm{e}} \rho_{\mathrm{e}} = -\rho_{\mathrm{e}}^{2} r_{\mathrm{e}} \boldsymbol{u}_{\mathrm{e}} \tag{3.41}$$

The friction force on the charges, f_{e}, is proportional to the drift velocity, u_{e}, with a volumetric force unit, $\mathrm{N/m^3}$. In this sense, the general entropy production for electrical conduction is given by

$$\sigma^{s} = -\frac{1}{T} \boldsymbol{f}_{\mathrm{e}} \cdot \boldsymbol{u}_{\mathrm{e}} = \frac{1}{T} \rho_{\mathrm{e}}^{2} r_{\mathrm{e}} \boldsymbol{u}_{\mathrm{e}} \cdot \boldsymbol{u}_{\mathrm{e}} = \frac{1}{T} r_{\mathrm{e}} \boldsymbol{i}^{2} \tag{3.42}$$

It is compatible with the Joule's first law. Hence the entropy production is the dissipation rate of electrical energy into the heat energy, divided by T. When the Ohm's law holds, Eq. 3.42 is equivalent to the traditional entropy production, Eq. 3.39. In fact, the Ohm's law actually omits the time needed for the acceleration of charge. The mass of charge is generally small, so the relaxation behavior in the electrical conduction is not easily observed. However, the Ohm's law can be deviated in extreme conditions. Havemann et al. [21] elucidated that the Ohm's law should be modified with nonlinear correction terms at very low temperature based on a derivation from the Boltzmann equation. A more obvious example is the ballistic transport inside the cathode ray tube where the inertia of electrons overwhelms friction effects, and the current direction could be opposite to the electrical field. If the electrical current and field in this case are inserted into Eq. 3.39, the entropy production is negative. If the general expression of entropy production, Eq. 3.42, is adopted, one has the right results that the real entropy is zero, because there is no dissipation in this process. In the analysis of thermoelectric effect based on EIT theory, the perspective is similar that the current, i, is regarded as an independent state variable of the system and written into the Gibbs equation.

3.5 Momentum Transport

The classical entropy production for the momentum transport process is

$$\sigma^{s} = -\frac{1}{T} \mathbf{P}^{v} : \nabla \boldsymbol{u}_{\mathrm{f}} \tag{3.43}$$

This equation only persist semi-positive definite when the Newton's law of viscosity holds. However, in many fluids, such as the colloids, polymer melts and solutions, the momentum transports don't obey the Newton's viscosity law. These fluids are called the non-Newtonian fluid, wherein the momentum transport is called the non-Newtonian flow [22–26]. The simplest non-Newtonian constitutive equation is the Maxwell model

$$\tau \frac{\mathfrak{D}\mathbf{P}^{v}}{\mathfrak{D}t} + \mathbf{P}^{v} = 2\eta \nabla \boldsymbol{u}_{\mathrm{f}} \tag{3.44}$$

where τ is the relaxation time, $\mathfrak{D}$ is the objective derivative, η is the viscosity coefficient. In the non-Newtonian flow, the classical expression for entropy production, Eq. 3.43, cannot maintain positive definite. In analogy to the analysis on other transport processes, $\mathbf{P}^{v}$ can be regarded as a momentum flux and proportional to a certain friction force. In the Newton's law of viscosity, the velocity gradient, ∇u_{f}, drives the momentum transport. It should correspond to the driving force. Thus the entropy production can be reformed as

$$\sigma^{s} = \frac{1}{T}\frac{1}{2\eta}\mathbf{P}^{v}:\mathbf{P}^{v} \tag{3.45}$$

This expression agrees with the EIT theory.

The drift velocity of the momentum transport is a bit more complex compared with other transport processes. Consider the momentum transport in a constant volume system, such as the steady two dimensional plate Couette flow. In this simplest flow pattern, the momentum flux, as well as the shear stress, is constant across each plane parallel to the velocity direction In this case the drift velocity can be defined as

$$\boldsymbol{u}_{\mathrm{m}} = \frac{\mathbf{P}^{v} \cdot \boldsymbol{u}_{\mathrm{f}}}{\rho \boldsymbol{u}_{\mathrm{f}}^{2}} \tag{3.46}$$

where u_{m} is the momentum transport velocity. The numerator in Eq. 3.46 represents the flux of shear work, which is in the same unit of heat flux. The denominator is the local density of kinetic energy, which is in the same unit of thermal energy. Therefore, Eq. 3.46 has the similar definition to the drift velocity of thermomass, u_{h}. u_{m} can be regarded as the drift velocity of the mechanical energy of a fluid.

In the simple plate Couette flow, the normal stress is uniform in the whole region. Thus the Newton's viscosity law can be written as

$$P_{xy}^{v} = -2\eta \frac{\partial}{\partial y} u_{\mathrm{f}x} \tag{3.47}$$

Here P_{xy}^{v} can be understood as not only the shear stress in the xy plane, but also the momentum flux density across the horizontal plane. In the picture of shear stress Eq. 3.47 is the constitutive equation of the stress and strain. In the picture of momentum flux, Eq. 3.47 turns to the linear phenomenological law for momentum transport. Multiplied with the fluid velocity u_{fx}, Eq. 3.47 becomes

$$P^{v}_{xy}u_{\mathrm{fx}} = -2\frac{\eta}{\rho}\frac{\partial}{\partial y}\rho u^{2}_{\mathrm{f}x} \tag{3.48}$$

This equation corresponds to the linear phenomenological law for the transport of kinetic energy. The driving force in Eq. 3.48 is the gradient of the density of the kinetic energy, which is in analogy to the temperature gradient (the gradient of internal energy density) in the Fourier's heat conduction law. Kay and Nedderman [27] analyzed the physics of Navier-Stokes equations and indicated that the $\mathbf{P}^{\mathrm{v}}\cdot\boldsymbol{u}_{\mathrm{f}}$ represents the power rate of the shear work, namely the rate of diffusive transport of the mechanical energy through the shear stress.

In a transient Couette flow between plates, the conservation of momentum gives

$$\frac{\partial \rho u_{\mathrm{f}x}}{\partial t} = -\frac{\partial}{\partial y}P^{v}_{xy} \tag{3.49}$$

Combination of Eqs. 3.49 to 3.47 leads to

$$\frac{\partial \rho u_{\mathrm{f}x}}{\partial t} = 2\eta\frac{\partial^{2}}{\partial y^{2}}u_{\mathrm{f}x} \tag{3.50}$$

which is a diffusive type equation. The disturbance of the velocity propagates with an infinite speed based on Eq. 3.50. Actually, the momentum diffusion in a fluid also depends on the thermal motion of molecular. Thus the propagation speed of velocity disturbance should be in the same magnitude of the speed of thermal disturbance. Usually only the compression-expansion wave is discussed in fluids. The shear wave is rarely noticed. The reason could be that the fluid element cannot sustain the translational shear stress. Its deformation causes the strong dispersion of wave, making it difficult to observe the shear wave in fluids. Nevertheless, in ultrafast shearing or X-ray/neutron scattering processes, the impact time is so short that the fluid element doesn't have enough time to dissipate the deformation work through the thermal motion. Therefore, the fluid will partly exhibit the solid-like feature, namely, becomes elastic to some extent. For this reason, the constitutive equations with the relaxation term, such as Eq. 3.44, are called the viscoelastic constitutive equations. Similarly, the CV model for heat conduction is also called the viscoelastic heat conduction model in some literature.

Combining Eq. 3.49 with the Maxwell model, Eq. 3.44, gives

$$\tau\frac{\partial^{2}\rho u_{\mathrm{f}x}}{\partial t^{2}} + \frac{\partial \rho u_{\mathrm{f}x}}{\partial t} = 2\eta\frac{\partial^{2}}{\partial y^{2}}u_{\mathrm{f}x} \tag{3.51}$$

Note that the objective derivative $\mathfrak{D}/\mathfrak{D}t$ can be simplified as the partial differential derivative in the case of simple Couette flow. Equation 3.51 is a hyperbolic type equation. It predicts the finite propagation speed of velocity disturbance as

$$v_{\mathrm{sf}} = \sqrt{\frac{2\eta}{\rho\tau}} \tag{3.52}$$

The shear modulus can be defined as

$$G_{\mathrm{f}} = \frac{2\eta}{\tau} \tag{3.53}$$

Thus Eq. 3.52 turns to

$$v_{\mathrm{sf}} = \sqrt{\frac{G_{\mathrm{f}}}{\rho}} \tag{3.54}$$

This is actually the same form as the shear wave speed in solids. In the expression of G_{f}, the dynamic viscosity η represents the degree of thermal motion of fluid molecular, while the relaxation time τ characterizes the inertia of fluid molecular, i.e. the ability of resisting deformation. The relaxation time is generally larger for fluids with larger viscosity. The shear wave speeds of fluids at room temperature are in the range of several hundred m/s, which is close to the thermal motion speed of gases at the same temperature. The relaxation time of some fluids can be extracted from experiments, as shown in Table 3.1 [3]. v is the dynamic viscosity, 2 η/ρ, v_{L} is the speed of the longitudinal wave in the fluid. For non-polar molecules such as the carbon tetrachloride and benzene, the shear wave speeds can be obtained from Table 3.1, which are 481 and 665 m/s, respectively. The shear wave speeds are about half of their longitudinal wave speed. This relation is similar to case in solids. The molecular listed in Table 3.1 are all small molecular, with the relaxation times in the magnitude of 10^{-12}s. Therefore, the non-Newtonian behavior of these fluids is only detectable in ultrafast shear and measurement. For fluids of large molecular, such as the polymer melts, the relaxation time can be as large as 0.01 s. The non-Newtonian behavior can be measured in ordinary conditions. However, the

Table 3.1 Properties of some liquids [3]

	$10^{12}\tau$ (s)	$10^4\eta$ (Ns/m^2)	10^7 ν (m^2/s)	v_L (m/s)
Carbon tetrachloride	2.46	4.85	6.11	926
Chloroform	2.08	2.90	3.91	995
Carbon disulfide	1.38	1.82	3.3	1149
Benzene	1.67	3.26	7.40	1298
Toluene	1.60	2.95	6.8	1275
Acetone	2.19	1.60	4.1	1174

flow of large molecular fluid may include the change of geometry conformation and the energy conversion. Thus their behavior cannot be simply described by the Maxwell model.

Comparing Eq. 3.44 with the general heat conduction law, Eq. 2.20, one concludes that the relaxation term in Eq. 3.44 corresponds to the inertia term in the momentum (or kinetic energy) transport. The momentum flux, $\mathbf{P}^v$, corresponds to the friction force. In analogy to the thermomass theory, the friction force during the momentum transport can be defined as

$$\boldsymbol{f}_{\mathrm{m}} = -\frac{\mathbf{P}^v \cdot \rho \boldsymbol{u}_{\mathrm{f}}}{\eta} = -\frac{(\rho \boldsymbol{u}_{\mathrm{f}})^2}{\eta} \boldsymbol{u}_{\mathrm{m}} = -\frac{\rho \boldsymbol{u}_f^2}{\eta/\rho} \boldsymbol{u}_{\mathrm{m}} \tag{3.55}$$

This expression is similar to Eqs. 3.32 and 3.33. The dynamic viscosity η/ρ has a unit of m^2/s. $\rho \boldsymbol{u}_{\mathrm{f}}^2$ correspond to p_k and p_{h}, with a unit of the energy density. However, this expression is developed for the simplest case of flow pattern. For more general cases, the momentum transport will induce the variation of the static/normal pressure and gravitational potential, which means the kinetic energy of fluid can convert into the potential of gravitational energy. In these cases the drift velocity cannot be expressed simply as Eq. 3.46.

In an extreme case that the friction force in the translational momentum transport is zero, the mechanical energy doesn't dissipate. Thus Eq. 3.44 transforms as

$$\tau \frac{\mathfrak{D}\mathbf{P}^v}{\mathfrak{D}t} = 2\eta \nabla \boldsymbol{u}_{\mathrm{f}} \tag{3.56}$$

Combining it with the momentum balance equation, Eq. 3.49 gives

$$\frac{\partial^2 u_{\mathrm{f}x}}{\partial t^2} = \frac{2\eta}{\tau\rho} \frac{\partial^2 u_{\mathrm{f}x}}{\partial y^2} = v_{\mathrm{sf}}^2 \frac{\partial^2 u_{\mathrm{f}x}}{\partial y^2} \tag{3.57}$$

It is a non-damping wave equation. Substitution of Eq. 3.54 into Eq. 3.57 makes it the same form as the wave propagation equation in solid. Therefore, the constitutive equation, Eq. 3.56, indicates the material is elastic and satisfies the Hooke's law. Then the Newton's viscosity law is the pure diffusive transport of momentum while the Hooke's law is the pure ballistic transport.

3.6 Conclusion

The entropy production in the classical irreversible thermodynamics is the bilinear product of generalized forces and fluxes. It is only applicable in the condition of linear transports. When the linear transport laws are invalid, such as the non-Fourier heat conduction, the classical energy production could be negative, which violates the second law of thermodynamics. EIT defines the additional independent state variables, such as the heat flux. In this way the expression of classical energy production is modified to be compatible with the non-Fourier heat conduction.

The dynamical analysis based on the thermomass theory indicates that the Fourier's law of heat conduction is the balance between the friction and driving forces. In non-Fourier heat conduction, the driving force is unbalanced with the friction force. The entropy measures the irreversibility. It should be proportional to the product of the friction force and the thermomass drift velocity, which leads to the general expression of entropy production. The general entropy production is the quadratic function of heat flux. It is suitable for both the ordinary Fourier heat conduction and the non-Fourier heat conduction in extreme conditions.

The driving forces, friction forces and the drift velocity can be also defined in other transport processes, such as the mass diffusion, electrical conduction and the momentum transport. The general entropy production should be expressed as the minus friction force multiplied by the drift velocity, then divided by T. The friction force is always opposite to the drift velocity, which ensures the semi-positive definite of entropy production. When the linear transport laws are applicable, the driving force balances the friction force, and the general entropy production is equivalent to the classical one. When the inertia forces are non-negligible, the friction force and driving force are unbalanced. In this case only the general entropy production can evaluate the irreversibility of transport processes correctly.

The combined expression for the general entropy production for the above mentioned transport processes can be written as

$$\sigma^s = \frac{1}{\kappa T^2}\boldsymbol{q}\cdot\boldsymbol{q} + \frac{1}{2\eta T}\mathbf{P}^v : \mathbf{P}^v + \frac{1}{T}\sum_{k=1}^{N}\left(\boldsymbol{u}_k\cdot\sum_{j=1}^{N}\Lambda_{kj}\boldsymbol{u}_j\right) + \frac{1}{T}r_e\boldsymbol{i}\cdot\boldsymbol{i} \tag{3.58}$$

In Table 3.2 the comparison of the classical and general entropy productions is presented.

Table 3.2 Classical and general entropy productions in transports

Irreversible processes		Vectorial			Tensorial
		Heat conduction	Mass diffusion	Electrical conduction	Viscous flow
Classical $\sigma^s = \boldsymbol{J}\cdot\boldsymbol{X}$	σ^s	$\boldsymbol{q}\cdot\nabla\left(\frac{1}{T}\right)$	$-\sum_{k=1}^{N}\boldsymbol{J}_k\cdot\nabla\left(\frac{\mu_k}{T}\right)$	$-\frac{1}{T}\boldsymbol{i}\cdot\nabla\varphi_e$	$-\frac{1}{T}\mathbf{P}^v{:}\nabla\boldsymbol{v}$
	$\boldsymbol{J}$	$\boldsymbol{q}$	$\boldsymbol{J}_k$	$\boldsymbol{i}$	$\mathbf{P}^v$
	$\boldsymbol{X}$	$\nabla\left(\frac{1}{T}\right)$	$-\nabla\left(\frac{\mu_k}{T}\right)$	$-\frac{1}{T}\nabla\varphi_e$	$-\frac{1}{T}\nabla\mathbf{v}$
General $\sigma^s = \frac{\boldsymbol{u}\cdot\boldsymbol{f}}{T}$	σ^s	$\frac{1}{\kappa T^2}\boldsymbol{q}\cdot\boldsymbol{q}$	$\frac{1}{T}\sum_{k=1}^{N}\left(\boldsymbol{u}_k\cdot\sum_{j=1}^{N}\Lambda_{kj}\boldsymbol{u}_j\right)$	$\frac{1}{T}\rho_e^2 r_e\boldsymbol{u}_e\cdot\boldsymbol{u}_e$	$\frac{1}{T}\frac{1}{2\eta}\mathbf{P}^v{:}\mathbf{P}^v$
	$\boldsymbol{u}$	$\boldsymbol{u}_h = \frac{\boldsymbol{q}}{\rho C_V T}$	$\boldsymbol{u}_k = \boldsymbol{J}_k/\rho_k$	$\boldsymbol{u}_e = \boldsymbol{i}/\rho_e$	$\boldsymbol{u}_m = \frac{\mathbf{P}^v\cdot\boldsymbol{u}_f}{\rho\boldsymbol{u}_f^2}$
	$\boldsymbol{f}$	$-\frac{(\rho C_V)^2 T}{\kappa}\boldsymbol{u}_h$	$-\sum_{j=1}^{N}\Lambda_{kj}\boldsymbol{u}_j$	$-\rho_e^2 r_e\boldsymbol{u}_e$	$-\frac{\rho\boldsymbol{u}_f^2}{\eta/\rho}\boldsymbol{u}_m$

References

1. Jou D, Casas-Vazquez J, Lebon G (1999) Extended irreversible thermodynamics revisited (1988–98). Rep Prog Phys 62(7):1035
2. Lebon G, Jou D, Casas-Vázquez J (2008) Understanding non-equilibrium thermodynamics. Springer, Berlin
3. Jou D, Casas-Vázquez J, Lebon G (2010) Extended irreversible thermodynamics. Springer, Berlin
4. Cimmelli VA (2009) Different thermodynamic theories and different heat conduction laws. J Non-Equilib Thermodyn 34(4):299–333
5. Criado-Sancho M, Llebot JE (1993) Behavior of entropy in hyperbolic heat conduction. Phys Rev E 47:4104–4107
6. Casas-Vázquez J, Jou D (1994) Nonequilibrium temperature versus local-equilibrium temperature. Phys Rev E 49(2):1040
7. Casas-Vazquez J, Jou D (2003) Temperature in non-equilibrium states: a review of open problems and current proposals. Rep Prog Phys 66(11):1937
8. Cimmelli VA, Sellitto A, Jou D (2009) Nonlocal effects and second sound in a nonequilibrium steady state. Phys Rev B 79(1):014303
9. Cimmelli VA, Sellitto A, Jou D (2010) Nonequilibrium temperatures, heat waves, and nonlinear heat transport equations. Phys Rev B 81(5):054301
10. Cimmelli VA, Sellitto A, Jou D (2010) Nonlinear evolution and stability of the heat flow in nanosystems: beyond linear phonon hydrodynamics. Phys Rev B 82(18):184302
11. Alvarez FX, Jou D (2008) Size and frequency dependence of effective thermal conductivity in nanosystems. J Appl Phys 103(9):094321
12. Sellitto A, Alvarez FX, Jou D (2010) Second law of thermodynamics and phonon-boundary conditions in nanowires. J Appl Phys 107(6):064302
13. Jou D, Criado-Sancho M, Casas-Vázquez J (2010) Heat fluctuations and phonon hydrodynamics in nanowires. J Appl Phys 107(8):084302
14. Jou D, Cimmelli VA, Sellitto A (2012) Nonlocal heat transport with phonons and electrons: application to metallic nanowires. Int J Heat Mass Transf 55(9):2338–2344
15. Sellitto A, Alvarez FX, Jou D (2012) Geometrical dependence of thermal conductivity in elliptical and rectangular nanowires. Int J Heat Mass Transf 55(11):3114–3120
16. Sellitto A, Cimmelli VA, Jou D (2013) Entropy flux and anomalous axial heat transport at the nanoscale. Phys Rev B 87(5):054302
17. Mazur P, de Groot SR (1963) Non-equilibrium thermodynamics. North-Holland, Amsterdam
18. Vignes A (1966) Diffusion in binary solutions. Variation of diffusion coefficient with composition. Ind Eng Chem Fundam 5(2):189–199
19. Atkins P, Paula J (2006) Physical chemistry, 8th edn. Oxford University Press, New York
20. Gallavotti G (1996) Chaotic hypothesis: onsager reciprocity and fluctuation-dissipation theorem. J Stat Phys 84(5–6):899–925
21. Havemann RH, Engel PF, Baird JR (2003) Nonlinear correction to Ohm's law derived from Boltzmann's equation. Appl Phys Lett 24(8):362–364
22. Bird R B, Armstrong R C, Hassager O (1987) Dynamics of polymeric liquids, 2nd edn. Wiley, New York
23. Barnes HA, Hutton JF, Walters K (1989) An introduction to rheology. Elsevier, New York
24. Evans DJ(1988) Rheological properties of simple fluids by computer simulation. Phys Rev A, 23(4):1981
25. Erpenbeck JJ (1984) Shear viscosity of the hard-sphere fluid via nonequilibrium molecular dynamics. Phys Rev Lett 52(15):1333
26. Yong X, Zhang LT (2012) Nanoscale simple-fluid behavior under steady shear. Phys Rev E 85 (5):051202
27. Kay JM, Nedderman RM (1985) Fluid mechanics and transfer processes. Cambridge University Press, Cambridge

Chapter 4
Nonequilibrium Temperature in Non-Fourier Heat Conduction

Abstract The definition of temperature is the foundation of thermodynamics. In extended irreversible thermodynamics (EIT) which goes beyond the classical one to be compatible with the non-Fourier heat conduction, the nonequilibrium temperature is defined. Based on the thermomass theory, it is shown that the static and stagnant pressures of the thermomass flow correspond to the static and stagnant temperatures, respectively. The stagnant temperature is higher than the static one due to the kinetic energy of thermomass. The static temperature is the real state variable, which is identical to the nonequilibrium temperature. It should be the criterion of thermodynamic equilibrium. The local entropy and internal energy densities should be represented by the static temperature. In this manner, the classical relation between entropy, internal energy, and temperature still holds. The derivation based on the phonon Boltzmann equation shows that the integral of the second-order expansion of the distribution function in the energy balance equation corresponds to the kinetic energy of thermomass, which leads to the difference between the static and stagnant temperatures.

Temperature is the basic quantity in physics. There are various ways to define temperature [1]. It can be based on the zeroth and second laws of classical thermodynamics, or the entropy and entropy flux of the thermodynamic equation of state, or the kinetic theory and the fluctuation theory in statistical physics. In equilibrium thermodynamics, the microscopic definition of temperature requires the principals of equipartition of energy and ergodicity. The macroscopic definitions are well consisted with the microscopic ones. In classical irreversible thermodynamics, the local equilibrium hypothesis is adopted which assumes the local microscopic distribution can be approximated by the equilibrium one. Therefore, the temperature can be still defined as in the equilibrium systems. However, when the degree of nonequilibrium becomes larger, the temperature could be different through different definitions. For example, the inverse of temperature equals the derivative of the entropy with respect to the internal energy in equilibrium thermodynamics. The expressions of entropy and entropy production are modified by the extended irreversible thermodynamics (EIT) to avoid the paradox of negative entropy production in non-Fourier heat conduction. In this condition the derivative

Y. Dong, *Dynamical Analysis of Non-Fourier Heat Conduction and Its Application in Nanosystems*, Springer Theses,
DOI 10.1007/978-3-662-48485-2_4

of entropy with respect to internal energy gives a temperature different from that in the equilibrium condition. The changed expression of temperature due to the modification of entropy is called the nonequilibrium temperature in EIT. Such difference is attributed to the effect of local nonequilibrium. In this chapter, the definition of temperature in non-Fourier heat conduction will be investigated from the thermomass perspective. Comparison with the nonequilibrium temperature defined by EIT will also be discussed.

4.1 Nonequilibrium Temperature in EIT

In EIT, the entropy is modified through introducing new state variables. In the heat conduction in nondeformable solids, the extended entropy is derived as [1–3]

$$s = s_{\mathrm{eq}} - \frac{1}{2}\frac{\tau}{\kappa T^2}\boldsymbol{q}\cdot\boldsymbol{q} = \rho C_{\mathrm{V}}\ln T - \frac{1}{2}\frac{\tau}{\kappa T^2}\boldsymbol{q}\cdot\boldsymbol{q} \tag{4.1}$$

where τ is the relaxation time based on the Cattaneo-Vernotte (CV) model. In equilibrium thermodynamics, the temperature relates to the entropy and internal energy as

$$T^{-1} = \frac{\partial s}{\partial e} \tag{4.2}$$

Incorporating Eq. 4.2 with Eq. 4.1 gives

$$\theta^{-1} = \frac{1}{T_{\mathrm{eq}}} + \frac{\tau}{\kappa\rho C_{\mathrm{V}} T_{\mathrm{eq}}^3}\boldsymbol{q}\cdot\boldsymbol{q} \tag{4.3}$$

where the first term of the right-hand side is the temperature in equilibrium situation, the second term is the modification term in case of the irreversible transport. Since the coefficient of the q^2 term is positive, the nonequilibrium temperature should be lower than the equilibrium one with the presence of heat flux. Assume τ is a constant, Eq. 4.3 can be transformed as

$$\theta = \frac{1}{T_{\mathrm{eq}}^{-1} + \tau\Big/\rho\kappa C_{\mathrm{V}} T_{\mathrm{eq}}^3\,\boldsymbol{q}\cdot\boldsymbol{q}} = T_{\mathrm{eq}}\left(1 - \frac{\tau}{\rho\kappa C_{\mathrm{V}} T_{\mathrm{eq}}^2}\boldsymbol{q}\cdot\boldsymbol{q} + o\left(q^2\right)\right) \tag{4.4}$$

The physical meaning of the nonequilibrium temperature has been investigated by EIT. From the perspective of the zeroth law it is regarded as the criterion of thermodynamic balance between systems. Consider a two-body system shown in Fig. 4.1 [1, 2]. The body A is at thermodynamic equilibrium with a temperature T_{eq}. The two ends of body B are at temperatures of $T_{\mathrm{eq}} + \Delta T$ and $T_{\mathrm{eq}} - \Delta T$, respectively.

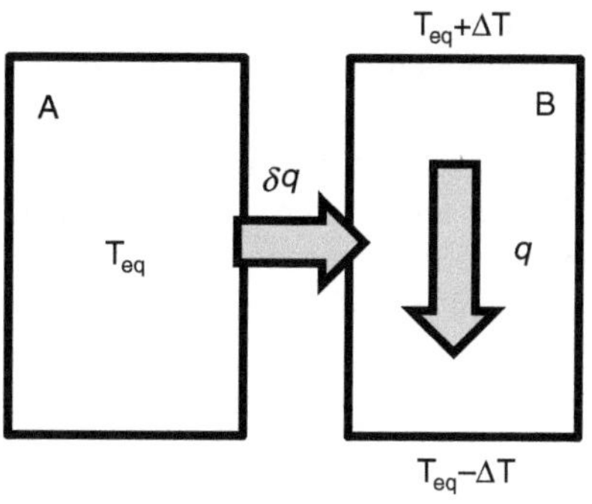

Fig. 4.1 Gedanken experiment to distinguish equilibrium and nonequilibrium temperatures [1]

The center of B is also at an equilibrium temperature, T_{eq}. However, the temperature gradient in body B induces a heat flux, q, making the nonequilibrium temperature lower than the equilibrium one. If the centers of A and B are connected by a thermal conductive medium, there would be a small heat flux, δq, passing through this medium. This gedanken experiment shows that heat transport can happen between two systems with the same T_{eq} but different θ. Therefore, the nonequilibrium temperature is the criterion of thermodynamic equilibrium.

In ordinary cases the difference between the equilibrium and nonequilibrium temperatures is quite small. For instances, for the CO_2 gas at room temperature and 0.1 atm, with a heat flux 10^9 W/m^2, the nonequilibrium temperature is 9.6×10^{-2} K lower than the equilibrium one [2]. For the single crystal silicon at room temperature, with a heat flux 10^{10} W/m^2, the nonequilibrium temperature is 0.2 K lower than the equilibrium one.

The nonequilibrium temperature in irreversible thermodynamics is also investigated by other researchers. Cimmelli et al. [4–6] proposed the dynamic nonequilibrium temperature, β, through introducing an additional relaxation term in the nonequilibrium temperature of EIT. This dynamic nonequilibrium temperature can be used to analyze the second sound propagation in solids. Baranyai [7] studied the operational temperature in the nonequilibrium molecular dynamics simulations. It was found that the operational temperature depends on the direction of perturbation on the system and deviates from the traditional temperature defined in equilibrium systems. Nevertheless, this difference is attributed to the algorithm of molecular dynamics simulations. Hatano and Jou [8] studied the temperature of the one-dimensional simple harmonic oscillator through a statistical perspective. They found that the temperature is different through different definitions and measurement methods. The configurational temperature and kinetic temperature are defined. The former represents the potential energy of system while the latter relates to the average kinetic energy of molecular, which is close to the classical definition of temperature in the kinetic theory. Furthermore, they elucidated that at higher dimensional systems the nonequilibrium temperature could be anisotropic.

In the classical nonequilibrium thermodynamics, the temperature is only well-defined as long as the local equilibrium hypothesis holds. If one does not assume the local nonequilibrium, the temperature is hard to define. In EIT's derivation, the equilibrium temperature is still used in the expression of nonequilibrium temperature. It is not clear enough that how to use the equilibrium temperature in the nonequilibrium conditions. The derivation of Eq. 4.1 is based on the transient non-Fourier heat conduction model, the CV model. However, the expression of θ does not contain the time derivative term. It should be applicable for both transient and steady conditions. One would desire to clarify that the definition of nonequilibrium temperature is independent of transport models, only relates to the nonequilibrium state of systems.

4.2 Zeroth Law

The zeroth law of thermodynamics indicates that if two bodies, A and B, are in thermodynamic equilibrium, and B and C are also in thermodynamic equilibrium, then A and C must be in equilibrium as well. Hence the systems in thermodynamic equilibrium construct a set with the same temperature. The zeroth law allows the temperature to be a measurable quantity. Based on the Fourier's law, the heat conduction only happens between systems with different temperatures with the heat flux proportional to the temperature gradient. Therefore, in the same conduction medium, the heat flux increases with the temperature difference.

In the thermomass theory the heat transport is regarded as the flow of a weighable, compressible fluid. The driving force on this flow is the gradient of the hydrostatic pressure. Consider two tanks filled with fluid and connected with a tube, the mechanical equilibrium is achieved when the hydrostatic pressure is the same at both ends of the tube. In this case the fluid will not flow from one tank to another. For the thermomass fluid, the criterion of mechanical equilibrium is also the hydrostatic pressure. The hydrostatic pressure of thermomass is expressed by Eq. 2.17, which contains not only the temperature but also the material properties such as density, specific heat and Grüneisen constant. However, Eq. 2.17 is actually the apparent pressure in the porous medium. For porous flow the mechanical equilibrium is determined by the intrinsic pressure which should remove the effect of porosity. The intrinsic pressure of thermomass should be independent of the material properties and only a function of temperature. For simplicity the variation of material properties is not considered here. Hence the apparent pressure, Eq. 2.17 can be used to investigate the balance and flow of thermomass.

For a compressible gas flow, one can define the static pressure and the stagnant pressure. The static pressure is the real pressure which does not depend on the speed of gas and the choice of reference framework. It is solely determined by the thermodynamic state of gases. The stagnant pressure (total pressure) is generally larger than the static pressure, with the difference called the dynamic pressure. The dynamic pressure is proportional to the kinetic energy of gas. The relation between the static pressure, p_{h}, and the stagnant pressure of phonon gas $p_{\mathrm{h,t}}$ can be established through the Bernoulli equation

$$\gamma\frac{p_{\mathrm{h}}}{\rho_{\mathrm{h}}}+\frac{1}{2}u_{\mathrm{h}}^{2}=\gamma\frac{p_{\mathrm{h,t}}}{\rho_{\mathrm{h}}} \tag{4.5}$$

where γ is the adiabatic index. In materials with uniform physical properties, the pressure and density of phonon gas satisfy $p_{\mathrm{h}}(\rho_{\mathrm{h}})^{-2}=\mathrm{const}$, hence $\gamma=2$ (cf. Eq. 2.59). In analogy to the static pressure, the stagnant phonon gas pressure can be defined as

$$p_{\mathrm{h,t}}=\frac{\gamma_{\mathrm{G}}\rho(C_{\mathrm{V}}T_{\mathrm{t}})^{2}}{c^{2}} \tag{4.6}$$

where T_{t} is the stagnant temperature, or total temperature, of phonon gas. The relation between the static temperature, T, and the stagnant temperature, T_{t}, of phonon gas can be extracted from Eqs. 4.5 and 4.6

$$T=T_{\mathrm{t}}-\frac{1}{4}\frac{q^{2}}{\gamma_{\mathrm{G}}C_{\mathrm{V}}(\rho C_{\mathrm{V}}T)^{2}}=T_{\mathrm{t}}-\frac{1}{2}\frac{\tau_{\mathrm{TM}}}{\kappa\rho C_{\mathrm{V}}T}q^{2} \tag{4.7}$$

This relation can be also written as

$$\frac{T_{\mathrm{t}}}{T}=1+\frac{\gamma-1}{2}\mathrm{Ma}_{\mathrm{h}}^{2} \tag{4.8}$$

where Ma_{h} is the Mach number of phonon gas. Equation 4.8 agrees well with the relation between the static temperature and stagnant temperature in compressible gas dynamics [9].

Equations 4.5 and 4.7 show that the static pressure of phonon gas is lower than the stagnant pressure during heat conduction. Since the phonon pressure directly relates to the temperature, the static temperature of phonon gas is lower than the stagnant temperature. The static pressure is the real pressure in gas dynamics. The gradient of static pressure drives the motion of gas. Therefore, the heat conduction happens as long as there is a difference of static temperature between two bodies. In other words, the static temperature is the criterion of thermodynamic equilibrium. Compared with EIT, the static temperature corresponds to the nonequilibrium temperature, θ, while the stagnant temperature corresponds to the equilibrium temperature, T_{eq}. T_{eq} is supposed to carry the entire energy of the medium, which is similar to the stagnant temperature in gas dynamics. When the heat flux is stagnated, the nonequilibrium temperature equals the equilibrium one, meanwhile the static temperature equals the stagnant one.

From Eq. 4.5 one can see that the difference between the static and stagnant temperature originates from the kinetic energy of phonon gas. With the same stagnant temperature, a larger kinetic energy of phonon gas, i.e., the heat flux density, corresponds to the lower static temperature. In the nonviscous gas flow, the stagnation of velocity is reversible. Consider a reversible stagnation of a steady phonon gas flow, it can be extracted from Eq. 2.20 that

$$\begin{gathered}\rho_{\mathrm{h}} u_{\mathrm{h}} \cdot \nabla u_{\mathrm{h}} = -\nabla p_{\mathrm{h}} \\ \nabla \left[\frac{1}{2} \left(\frac{q}{\rho C_{\mathrm{V}} T} \right)^2 \right] = -2\gamma_{\mathrm{G}} C_{\mathrm{V}} \nabla T\end{gathered} \tag{4.9}$$

By integration one has

$$\nabla T_{\mathrm{t}} = \nabla \left[\gamma_{\mathrm{G}} C_{\mathrm{V}} T + \frac{1}{4} \left(\frac{q}{\rho C_{\mathrm{V}} T} \right)^2 \right] = 0 \tag{4.10}$$

Equation 4.10 indicates that the stagnant temperature keeps constant during the reversible process. It is in essence the conservation of kinetic and potential energy of phonon gas. The potential and kinetic energies can convert to each other in reversible processes, changing the static temperature of phonon gas. It is worth noticing that there is a slight difference between Eqs. 4.7 and 4.4. The coefficient of the q^2 term is 1/2 in the former while is unity in the latter. This difference will be addressed in the following sections.

For one-dimensional steady heat conduction, the heat flux is constant in the whole system. Thus the momentum balance equation of thermomass is simplified as

$$\begin{aligned}\boldsymbol{q} &= -\kappa(1-b)\nabla T \\ &= -\kappa \nabla \left(T + \frac{1}{2} \frac{\tau_{\mathrm{TM}}}{\kappa \rho C_{\mathrm{V}} T} q^2 \right) + o(q^2) \\ &= -\kappa \nabla T_{\mathrm{t}} + o(q^2)\end{aligned} \tag{4.11}$$

which indicates that the heat flux in the system can reduce the effective thermal conductivity. The reason for this behavior is that the friction on the phonon gas will decrease its pressure, which means the temperature is also decreased. The pressure loss will decrease the density since the phonon gas is a compressible fluid. The mass flux of phonon gas is constant. Therefore, the velocity of phonon gas increases along the heat conduction direction. The increased velocity corresponds to the decreased static pressure which drives the motion of phonon gas. Because of the loss of the effective driving force, the effective thermal conductivity decreases. This behavior is in analogy to the compressible gas flow in micro-channels, where the pressure loss due to viscous friction also causes the acceleration of gases.

Cimmelli et al. [4–6] proposed the dynamical nonequilibrium temperature to describe the non-Fourier heat conduction with both the transient nonlinear and the spatial nonlocal effects. The nonequilibrium temperature, θ, in EIT is only a function of heat flux and independent of the time evolution rate. The dynamical nonequilibrium temperature, β, connects to θ with a relaxation term

$$\tau \frac{\partial}{\partial t}\beta + \beta = \theta \tag{4.12}$$

This equation is similar to the CV model. At steady states, β is identical to θ. In nonsteady cases, β deviates from θ, and relaxes back to θ with a time constant, τ. In this way, β considers additionally the relaxational effect in transient cases. Cimmelli et al. [4–6] defined the heat flux through the Fourier's law with β

$$-\kappa \nabla \beta = q \tag{4.13}$$

The time derivative of Eq. 4.13 gives the CV model.

Based on the dynamical nonequilibrium temperature one can get the more general heat conduction models. Cimmelli et al. [4–6] introduced the gradient of β, $\beta_{,i} = \nabla_i \beta$, as the new independent state variables. Assume that the entropy does not depend on the gradient of internal energy, the extended entropy of a nonequilibrium system is obtained through the derivation based on the extended Liu procedure

$$s(e;\beta;\beta_{,k}) = s_0(e;\beta) - \frac{1}{2} s_\beta(e;\beta)\beta_{,i}\beta_{,i} \tag{4.14}$$

According to Eq. 4.13, $\beta_{,i}$ is proportional to the heat flux. Then the modification term in Eq. 4.14 is proportional to q^2, which is similar to the extended entropy in EIT. Here s_β is a variable coefficient. The entropy of system should be max at thermodynamic equilibrium and any nonequilibrium flux should decrease the system entropy. Therefore s_β is positive. The derivative of entropy with respect to $\beta_{,i}$ is

$$\lambda_i^{(\beta)} = \frac{\partial s}{\partial \beta_{,i}} = -s_\beta(e;\beta)\beta_{,i} \tag{4.15}$$

Assume that the time derivative of β has the form

$$\dot{\beta} = f(e;\beta;e_{,k};\beta_{,k}) \tag{4.16}$$

Its derivative with respect to $\beta_{,i}$ is

$$\frac{\partial f}{\partial \beta_{,k}} = \frac{1}{3\lambda_i^{(\beta)}} \frac{1}{\theta^2} q_i \frac{\partial \theta}{\partial \beta_{,k}} \tag{4.17}$$

Integration of Eq. 4.17 gives

$$f(e;\beta;\beta_{,k}) = f_0(e;\beta) + \frac{1}{2} f_1(e;\beta)\beta_{,i}\beta_{,i} \tag{4.18}$$

where f_1 is the derivative of s_β with respect to the internal energy, e

$$f_1 = \frac{\kappa}{s_\beta(e;\beta)} \frac{\partial s_\beta}{\partial e} \tag{4.19}$$

In case of the small heat flux, Eq. 4.18 should be able to recover Eq. 4.12, so

$$f_0(e;\beta) = -\frac{1}{\tau}(\beta - \theta) \tag{4.20}$$

Making a time derivative of Eq. 4.18 then inserting Eq. 4.13 yields

$$\tau \dot{q}_i + q_i = -\kappa \theta_{,i} - \frac{\tau f_1}{\kappa} q_k q_{k,i} \tag{4.21}$$

Therefore, one gets a general heat conduction model with the nonlocal effect. Compared with the model based on thermomass theory, Eqs. 2.20, 4.21 also contains a spatial convection term.

Compare Eq. 4.14 with Eq. 4.1 one can extract that

$$s_\beta = \frac{\tau\kappa}{T^2} \tag{4.22}$$

Inserting it in Eq. 4.19 leads to

$$f_1 = \frac{2\kappa}{\rho C_V T} \tag{4.23}$$

Substitution of Eqs. 4.20 and 4.23 into Eq. 4.18 gives

$$\dot{\beta} = -\frac{1}{\tau}(\beta - \theta) + \frac{q^2}{\kappa \rho C_V T} \tag{4.24}$$

The second term of the right-hand side of Eq. 4.24 is called the nonlocal term. Actually it is still "local" in the term of heat flux. Nevertheless, the heat flux in Eq. 4.24 is defined through the spatial gradient of β, so Eq. 4.24 is regarded as nonlocal with a dependence of both the time derivative of β and the gradient of β. The gradient of Eq. 4.24 is

$$\tau \frac{\mathrm{D}q}{\mathrm{D}t} + q = -\kappa\left(1 - \frac{\tau q^2}{\kappa \rho C_V T^2}\right)\nabla\theta \tag{4.25}$$

If θ is equivalent to T, Eq. 4.25 has the same form as the general heat conduction law, Eq. 2.20. The spatial convection term is included in the material derivative, D/Dt, which predicts the different propagation speed of heat wave in different directions. The nonlinear coefficient on the right hand of Eq. 4.25, $(1 - \tau q^2/\kappa \rho C_V T^2)$, comes from the q^2 term in Eq. 4.24, which predicts the decrease of effective thermal conductivity in the presence of large heat flux in the system. In steady conditions, β is equivalent to the stagnant temperature of phonon gas, T_t. Then Eq. 4.13 is the same as Eq. 4.11. Thus in steady states the Fourier's law can still hold with the representation of the stagnant temperature. The reason for such behavior is that the Fourier's law can be regarded as not only a transport model but also a dissipation model. The role of friction is to dissipate the total pressure of phonon gas, i.e., the total mechanical energy. Integration of Eq. 4.11 with respect to x gives

$$\frac{\boldsymbol{q}l}{\kappa} = T_t(x=0) - T_t(x=l) \tag{4.26}$$

Remind that q is proportional to the friction on phonon gas, the left-hand side of Eq. 4.26 represents the negative work done by the friction during the motion of phonon gas from $x = 0$ to $x = l$. The right-hand side is the dissipation of the mechanical energy of phonon gas. Therefore, the Fourier's law not only indicates the linear relation between the heat flux and temperature gradient, but also represents that the mechanical energy dissipation rate equals the negative power of friction force. The second picture implies that the temperature in the Fourier's law should be the stagnant temperature since it measures the total mechanical energy of phonon gas.

4.3 Second Law

In Sect. 3.2 the general entropy production is derived as

$$\sigma^s = \frac{q^2}{\kappa T^2} \tag{3.19}$$

Thus the expression of entropy should be modified as well. One has

$$\sigma^s = \frac{\mathrm{d}s}{\mathrm{d}t} + \nabla \cdot \boldsymbol{J}_s \tag{4.27}$$

The entropy flux is

$$\boldsymbol{J}_s = \frac{q}{T} \tag{4.28}$$

From Eqs. 3.19, 4.27 and 4.28 one obtains

$$\frac{\mathrm{d}s}{\mathrm{d}t} = -\nabla \cdot \boldsymbol{J}_s + \sigma^s = -\nabla \cdot \left(\frac{\boldsymbol{q}}{T}\right) + \sigma^{\mathrm{s}}_{\mathrm{TM}} = \frac{\boldsymbol{q}}{\kappa T^2} \cdot (\boldsymbol{q} + \kappa \nabla T) - \frac{\nabla \cdot \boldsymbol{q}}{T} \tag{4.29}$$

Note that the entropy flux still uses the expression in classical nonequilibrium thermodynamics. Müller [10] pointed out that the entropy flux of the multipolar material needs to be modified as

$$\boldsymbol{J}_s = \frac{\boldsymbol{q}}{T} + \mathcal{K} \tag{4.30}$$

where $\mathcal{K}$ is the modification factor relating to the polarity of material. Nevertheless, for most uniform materials, the classical expression, Eq. 4.28, is applicable. Sellitto et al. [11] proposed that in the case of heat conduction with largely variable cross section, for example, the heat dissipation from a point heat source in graphene flacks, the temperature at the intermediate part could be higher than that at the heat source. Thus they modified the heat flux as

$$\boldsymbol{J}_s = \frac{\boldsymbol{q}}{T} + \frac{\lambda^2}{\kappa T^2} \nabla \boldsymbol{q}^T \cdot \boldsymbol{q} \tag{4.31}$$

where λ is the average MFP of phonons. The modified entropy flux ensures the compatibility of this abnormal temperature distribution with the second law of thermodynamics. The discussion here does not consider the multipolar material and the heat conduction in nanosytems with large variation of cross section, thus the traditional entropy flux is still applicable.

In the case of small disturbance, the relaxation time of thermomass, τ_{TM}, can be assumed to be constant. Then the general heat conduction law, Eq. 2.20, can be written in a similar form to the CV model

$$\boldsymbol{q} + \kappa \nabla T = -\tau_{\mathrm{TM}} \frac{\partial \boldsymbol{q}}{\partial t} \tag{4.32}$$

The energy conservation relation is

$$\frac{\partial \rho C_{\mathrm{V}} T}{\partial t} + \nabla \cdot \boldsymbol{q} = 0 \tag{4.33}$$

Inserting Eqs. 4.32 and 4.33 into Eq. 4.29, and integrating in the time domain gives the general expression of entropy from the thermomass perspective

$$s = \rho C_{\mathrm{V}} \ln T - \frac{1}{2} \frac{\tau_{\mathrm{TM}}}{\kappa T^2} q^2 = s_{\mathrm{eq}} - \frac{1}{2} \frac{\tau_{\mathrm{TM}}}{\kappa T^2} q^2 \tag{4.34}$$

This expression is in consistent with the EIT expression, Eq. 4.1.

Consider the heat wave propagation in one dimensional isolated system described by Eq. 4.32, the local mechanical energy of thermomass fluid is

$$e_{\mathrm{h}} = e_{\mathrm{p}} + e_{\mathrm{k}} = \frac{\gamma_{\mathrm{G}} \rho C_{\mathrm{V}}^2 T^2}{c^2} + \frac{1}{2} \frac{q^2}{\rho C_{\mathrm{V}} T c^2} \tag{4.35}$$

The total mechanical energy of the system is

$$\begin{aligned} E_{\mathrm{h}} &= \int\limits_{\mathrm{V}} e_{\mathrm{h}} \mathrm{d}x = \int\limits_{\mathrm{V}} \left(\frac{\gamma_{\mathrm{G}} \rho C_{\mathrm{V}}^2 T^2}{c^2} + \frac{1}{2} \frac{q^2}{\rho C_{\mathrm{V}} T c^2} \right) \mathrm{d}x \\ &= \frac{\gamma_{\mathrm{G}} \rho C_{\mathrm{V}}^2}{c^2} \int\limits_{\mathrm{V}} \left(T^2 + \frac{1}{2} \frac{q^2}{\gamma_{\mathrm{G}} \rho^2 C_{\mathrm{V}}^3 T} \right) \mathrm{d}x \end{aligned} \tag{4.36}$$

The isolated system has no mass and energy exchange with other systems at the boundary, then the boundary heat flux is zero. One has

$$\int\limits_{\mathrm{V}} T \nabla \cdot \boldsymbol{q} \mathrm{d}x = -\int\limits_{\mathrm{V}} \nabla T \cdot \boldsymbol{q} \mathrm{d}x \tag{4.37}$$

Inserting Eq. 4.37 into Eq. 4.36 gives the time evolution of the mechanical energy of thermomass

$$\begin{aligned} \frac{\mathrm{d}E_{\mathrm{h}}}{\mathrm{d}t} &= \frac{\gamma_{\mathrm{G}} \rho C_{\mathrm{V}}^2}{c^2} \int\limits_{\mathrm{V}} \left(2T \frac{\partial T}{\partial t} + 2 \frac{\tau_{\mathrm{TM}}}{\kappa \rho C_{\mathrm{V}}} \frac{\partial q}{\partial t} \right) \mathrm{d}x \\ &= \frac{2 \gamma_{\mathrm{G}} C_{\mathrm{V}}}{\kappa c^2} \int\limits_{\mathrm{V}} (-q^2) \mathrm{d}x \end{aligned} \tag{4.38}$$

Integrating Eq. 4.34 in the space domain yields the total entropy of the system

$$S = \int\limits_{\mathrm{V}} s \mathrm{d}x = \int\limits_{\mathrm{V}} \left(\rho C_{\mathrm{V}} \ln T - \frac{1}{2} \frac{\tau_{\mathrm{TM}}}{\kappa T^2} q^2 \right) \mathrm{d}x \tag{4.39}$$

With the zero heat flux boundary, one has

$$\int\limits_{\mathrm{V}} \frac{\nabla \cdot \boldsymbol{q}}{T} \mathrm{d}x = \int\limits_{\mathrm{V}} \frac{\nabla T \cdot \boldsymbol{q}}{T^2} \mathrm{d}x \tag{4.40}$$

Then the time evolution of the general entropy of the system is

$$\begin{aligned} \frac{\mathrm{d}S}{\mathrm{d}t} &= \int\limits_{\mathrm{V}} \left(\frac{\rho C_{\mathrm{V}}}{T} \frac{\partial T}{\partial t} - \frac{\tau_{\mathrm{TM}} q}{\kappa T^2} \frac{\partial q}{\partial t} \right) \mathrm{d}x \\ &= \int\limits_{\mathrm{V}} \left(\frac{q^2}{\kappa T^2} \right) \mathrm{d}x \end{aligned} \tag{4.41}$$

Equations 4.38 and 4.41 show that the mechanical energy of thermomass keeps decreasing in this isolated system with internal heat wave propagation. Meanwhile the entropy keeps growing. The rate of mechanical energy dissipation and the entropy production are both proportional to q^2. Thus the heat flux in systems increases the rate of dissipation. With a zero heat flux, the rates of mechanical energy dissipation and entropy production both vanish.

Consider a system consisting of two small pieces of material with identical properties. The small size allows one to assume the constant temperature in each piece. At the initial state both pieces are at the temperature, T_0. Then a heat flux, q, is imposed between them, from A to B. In the reversible condition, the system will experience a temperature fluctuation. There is one moment when the heat flux in system is zero, and the temperatures of two pieces are $T_0 + \Delta T$ and $T_0 - \Delta T$, respectively. In this case the entropy of the system is

$$S = \rho V C_V \ln(T_0 + \Delta T) + \rho V C_V \ln(T_0 - \Delta T) \tag{4.42}$$

The entropy at the initial state based on the classical expression is

$$S_{0,\text{eq}} = 2V\rho C_V \ln T_0 \tag{4.43}$$

Apparently, the entropy in Eq. 4.43 is larger than that in Eq. 4.42. The distinction is expressed as

$$S_{0,\text{eq}} - S = 2V\rho C_V \ln T_0 - \rho C_V \ln\left(T_0^2 - \Delta T^2\right) = V\rho C_V \frac{\Delta T^2}{T_0^2} \tag{4.44}$$

If the heat conduction is reversible, the mechanical energy of thermomass is constant. The temperature fluctuation, ΔT, is obtained through Eq. 4.35

$$2\left[\gamma_G \rho (C_V T_0)^2 + \frac{1}{2}\frac{q^2}{\rho C_V T}\right] = \gamma_G \rho C_V^2 \left[(T_0 + \Delta T)^2 + (T_0 - \Delta T)^2\right] \tag{4.45}$$

Inserting Eq. 4.45 into Eq. 4.44 gives

$$\rho\left(s_{0,\text{eq}} - s\right) = \frac{1}{2}\rho C_V \frac{\Delta T^2}{T_0^2} = \frac{1}{2}\rho C_V \frac{q^2}{2\gamma_G \rho^2 C_V^3 T^3} = \frac{1}{2}\frac{\tau_{TM}}{\kappa T^2} q^2 \tag{4.46}$$

which is identical to Eq. 4.34. Compared with the derivation based on Eq. 4.29, the expression of entropy production is not used. Instead the derivation is based on the conservation of thermomass mechanical energy. A reversible stagnation process is assumed from the initial state to the zero heat flux state, which converts completely the kinetic energy of themomass to the potential energy. From Eq. 4.43 one also concludes that the entropy can be determined solely by the temperature only in the case of zero heat flux.

The heat flux can be fully determined by the local temperature gradient when the Fourier's law is applicable. Therefore, if the local temperature gradient is suddenly removed, the heat flux should be zero simultaneously. In non-Fourier heat conduction, the heat flux is decoupled with the temperature gradient. When the temperature gradient is suddenly removed, the heat flux can still exist and attenuate due to friction. In EIT, the remaining heat flux without the temperature gradient is called the "uncompensated heat" [2]. Consider a heat conduction process, the temperature gradient is removed at $t = 0$, then the heat flux will relax to zero under the effect of friction force, accompanied with the entropy increase. Therefore, the entropy at $t = 0$ should be less than that at the final equilibrium state. For another example, consider a small piece of material which is in a system with steady heat conduction. The heat flux passing through this piece is q_0. At $t = 0$ this piece is suddenly isolated, which is equivalent to removing the temperature gradient and keeping it adiabatic. The heat flux in this piece will experience a free attenuation process

$$\tau \frac{\partial q}{\partial t} + q = 0 \tag{4.47}$$

This process is in essence the balance between the inertia force and the friction force. The evolution of heat flux is

$$q(t) = q_0 \exp\left(-\frac{t}{\tau}\right) \tag{4.48}$$

After a long enough time the heat flux will be zero with the system entropy reaching its maximum. Integrating the entropy production gives the entropy at the initial state

$$s_{\mathrm{eq}} - s_{t=0} = \int_0^\infty \sigma^s \mathrm{d}t = \int_0^\infty \frac{q_0^2 \exp(-2t/\tau)}{\kappa T^2} \mathrm{d}t = \frac{\tau q_0^2}{2\kappa T^2} \tag{4.49}$$

The system temperature does not change in this process. The entropy increases due to the dissipation of the uncompensated heat. Equation 4.49 shows that the system entropy is different with the same temperature but the different heat fluxes.

Regard the difference between the static temperature and the stagnant temperature, the energy conservation relation, Eq. 4.33, should be rewritten as

$$\rho C_{\mathrm{V}} \frac{\partial T_{\mathrm{t}}}{\partial t} = -\nabla \cdot \boldsymbol{q} \tag{4.50}$$

The stagnant temperature, T_{t}, measures the total energy of the local element. q measures the flux of total energy passing through each cross section. Inserting Eq. 4.50 into Eq. 4.29 and integrating gives

$$s = \rho C_V \ln T_t - \frac{1}{2}\frac{\tau_{TM}}{\kappa T^2} q^2 + o\left(q^2\right) \tag{4.51}$$

Compared with Eq. 4.34, the first term on the right-hand side of Eq. 4.51 is the stagnant temperature, T_t, rather than the static temperature, T. In the same manner, it is inferred that in EIT's derivation, the temperature in Eq. 4.1 should be replaced by T_{eq}. The first term on the right-hand side of Eq. 4.1 is called the equilibrium entropy. The equilibrium temperature also appears in the expression of nonequilibrium temperature, Eq. 4.3. If the equilibrium entropy is required to be related to the equilibrium temperature, then Eq. 4.1 is consistent with Eq. 4.51. The second term on the right-hand sider of Eq. 4.51 is proportional to the kinetic energy of thermomass fluid

$$\frac{1}{2}\frac{\tau_{TM}}{\kappa T^2} q^2 = \frac{1}{2}\rho_h u_h^2 \Big/ \xi T \tag{4.52}$$

It indicates that the kinetic energy of phonon gas stores additional available energy which implies a decrease of entropy. The kinetic energy of phonon gas is the local state variable. It is only determined by the magnitude of heat flux, and does not depend on the constitutive equation of heat conduction. In steady state heat conduction, the kinetic energy of phonon gas still exists. So the expression of entropy should be modified as well.

The internal energy is proportion to the temperature in classical thermodynamics

$$e = \rho C_V T \tag{4.53}$$

When the system is moving, the directional velocity will add on the velocity of the thermal motion of molecular. In the gas dynamics, the local internal energy is a function of static temperature and doesn't relate to the kinetic energy of the directional motion. Therefore, the local internal energy of medium during heat conduction should exclude the kinetic energy induced by the directional motion of phonon gas, which leads to an internal energy lower than the total energy

$$e = \rho C_V \left(T_t - \frac{1}{2}\frac{\tau_{TM}}{\kappa \rho C_V T} q^2 + o\left(q^2\right) \right) < e_t = \rho C_V T_t \tag{4.54}$$

Note that the general entropy production derived based on thermomass theory, Eq. 4.51 can be further transformed as

$$\begin{aligned} s &= \rho C_V \ln T_t - \frac{1}{2}\frac{\tau_{TM}}{\kappa T^2} q^2 \\ &= \rho C_V \ln \left[T_t \left(1 - \frac{1}{2}\frac{\tau_{TM}}{\rho C_V \kappa T^2} q^2 + o\left(q^2\right) \right) \right] = \ln e \end{aligned} \tag{4.55}$$

From Eq. 4.55 it is inferred that because the entropy is expressed by T_t rather than T, the local entropy in Eqs. 4.43 and 4.49 cannot be simply represented by temperature. In presence of heat fluxes, $T_t > T$, the entropy based on the classical expression is larger than the real entropy in the system. If the static temperature T is used in the expression, then the local entropy can still be expressed by a function of temperature. The relation between temperature, entropy, and internal energy is

$$T^{-1} = \mathrm{d}s/\mathrm{d}e = \mathrm{d}s/\rho C_V \mathrm{d}T \tag{4.56}$$

which is identical to the classical expression. If the internal energy is measured by the stagnant temperature, one has

$$\begin{aligned}\theta &= \mathrm{d}e_t/\mathrm{d}s \\ &= T\left(1 - \frac{1}{2}\frac{\tau_{\mathrm{TM}}}{\kappa \rho C_V T^2} q^2\right) \\ &= T_t\left(1 - \frac{\tau_{\mathrm{TM}}}{\kappa \rho C_V T^2} q^2\right) + o(q^2)\end{aligned} \tag{4.57}$$

which agrees with the EIT's derivation, Eq. 4.3. The relation between the static temperature, T, and the stagnant temperature, T_t, is similar to that between the nonequilibrium temperature, θ, and the equilibrium temperature, T_{eq}. However, the coefficient is different between Eqs. 4.7 and 4.3. The entropy is identical in terms of the static temperature and the nonequilibrium temperature. The difference comes from the expression of internal energy. In EIT theory, the internal energy is defined with the equilibrium temperature, T_{eq}, which includes the thermal motion energy and the kinetic energy of the directional motion of phonon gas. Based on the thermomass analysis, the internal energy should be defined by the static energy, T, which excludes the contribution of the kinetic energy of directional motion. Regard that the internal energy should be independent of reference framework, it is reasonable to exclude the kinetic energy of phonon gas. Therefore, the difference of coefficients in the expression of temperature comes from the different definition of internal energy. The expression based on the thermomass analysis, Eq. 4.7, is more accurate.

4.4 Phonon Boltzmann Derivation

In Chap. 2 the general heat conduction law based on thermomass theory is derived from the phonon Boltzmann equation. The driving force comes from the zeroth order expansion around f_D of the distribution function. The transient inertia is from the first order expansion, while the spatial inertia rises from the second-order expansion (see Sect. 2.3.1). Multiplying the phonon Boltzmann equation with $\hbar/c^2$ and integrating gives

$$\frac{\partial \int_{k}\left(f_0^n + f_{++}^n\right)\hbar\omega^n}{\partial t} + \nabla_j \int_{k} f_+^n \hbar\omega^n v_{kj}^n = 0 \tag{2.43}$$

In the derivation of Chap. 2 the contribution of f_{++} is neglected. Thus the first term on the left-hand side of Eq. 2.43 equals the time derivative of internal energy. In fact, the temperature in equilibrium system can be integrated from the equilibrium Planck distribution of phonon gas, i.e., f_0. Therefore, the integration of f_0 in Eq. 2.43 can be defined as a temperature. However, with the presence of f_{++}, this temperature only presents part of the energy in the control volume. f_{++} is induced by the drift motion of phonon gas. The magnitude of f_{++} increases with the drift velocity of phonon gas. Combined with the discussion in above sections, one can infer that f_0 corresponds to the static temperature while $f_0 + f_{++}$ corresponds to the stagnant temperature, T_t

$$\int_{k} f_0^n \hbar\omega^n = \rho C_V T < \rho C_V T_t = e_t \tag{4.58}$$

Hence the difference between the static and stagnant temperatures is

$$T = T_t - \frac{1}{\rho C_V} \int_{k} f_{++}^n \hbar\omega^n \tag{4.59}$$

Since f_{++} contains the term proportional to the square of the drift velocity of the phonon gas, the second term on the right-hand side of Eq. 4.59 is proportional to q^2. Integration of this term gives

$$\int_{k} f_{++}^n \hbar\omega^n = \int_{k} \hbar\omega^n \frac{1}{2}\frac{\partial^2 f_0^n}{\partial\omega^2}(\boldsymbol{k}\cdot\boldsymbol{u}_D)^2 = \int_{k} \frac{3\boldsymbol{u}_D^2}{v_s^2}\hbar\omega^n f_0^n \tag{4.60}$$

Guyer and Krumhansl [12] derived the average group velocity of phonon gas as

$$\frac{1}{v_s^2} = \frac{\tau\rho C_V}{3\kappa} \tag{4.61}$$

Inserting Eqs. 4.60 and 4.61 into Eq. 4.59 gives

$$T = T_t - \frac{9}{16}\frac{\tau}{\kappa}\frac{q^2}{\rho C_V T} + o\left(q^2\right) \tag{4.62}$$

The coefficient 9/16 in Eq. 4.62 is slightly larger than the coefficient, 1/2, in Eq. 4.7. According to the discussion of Eq. 2.63, this distinction comes from the Doppler eEffect of the phonon gas drift.

The derivation based on the phonon Boltzmann equation shows that the energy conservation equation describes the balance of the total energy. The total energy consists of an equilibrium part which is from f_0 and a nonequilibrium part which is from f_{++}. The first part is the real internal energy while the second corresponds to the kinetic energy of thermomass. This agrees with the conclusion of Eq. 4.7.

This phonon Boltzmann derivation can be regarded as the microscopic explanation of the temperature definition during the heat conduction in dielectric solids. For the heat conduction in gases, the equivalent mass of heat, thermon, is attached in each molecular. For metals, the heat conduction is mainly contributed by electrons. For semiconductors, the phonons and electrons are both important in heat conduction. From a more general perspective, the non-Fourier heat conduction can be the start point to analyze the nonequilibrium temperature microscopically. If a statistical analysis can give the non-Fourier heat conduction model, it should indicate the difference of the state distribution function from the equilibrium one. It is expected that this distinction induces not only the modification term in the non-Fourier heat conduction model, but also leads to the general form of nonequilibrium temperature, internal energy and entropy.

4.5 Conclusion

The thermodynamic equilibrium refers either the local equilibrium or the global one. The local equilibrium corresponds to the local equilibrium temperature, while the global equilibrium corresponds to the global temperature. The temperature in classical thermodynamics can be only defined in the equilibrium systems. This system is at a global equilibrium. The heat conduction requires the temperature gradient which apparently breaks the global equilibrium. However, the local equilibrium is necessary to define the temperature. Thus the degree of the deviation from equilibrium should be limited. The temperature in the Fourier's law and the internal energy density should be the local equilibrium temperature. When the system is far from equilibrium, the temperature cannot be well-defined.

EIT derives the nonequilibrium temperature. The derivation has some imperfections. First, the nonequilibrium temperature is derived mathematically from the expression of extended entropy with the physical meaning incompletely discussed. Second, the equilibrium temperature and nonequilibrium temperature are both used in the expression of internal energy and entropy, while their distinction is ambiguous. Third, the derivation of nonequilibrium temperature depends on the CV model which is lack of generality.

Based on the thermomass theory, the driving force on the thermomass flow should be the gradient of the local static pressure. The static pressure is always lower than the total pressure or the stagnant pressure in condition of heat conduction. The distinction is the dynamic pressure. The function of thermomass temperature and pressure gives the static temperature and stagnant temperature of thermomass. The static temperature, T, has the consistent definition to the nonequilibrium temperature, θ. The stagnant temperature, T_t, is defined as the equilibrium temperature, T_{eq}, in EIT. The static temperature is the real state variable of the system. It is independent of the moving speed, neither of the heat conduction processes. In the general heat conduction law, the gradient of static temperature is the real driving force. The stagnant temperature consists of the contribution from the static temperature and the kinetic energy of thermomass, which represents the total mechanical energy of thermomass. In one-dimensional steady flow, the Fourier's law defined with the stagnant temperature could have wider application region, which agrees with the derivation based on the dynamical nonequilibrium temperature. The modification term of the stagnant temperature is proportional to the square of heat flux, and the inverse of static temperature. Thus the effective thermal conductivity decreases with the increase of heat flux. This behavior can be understood from the dissipative feature of the Fourier's law.

In extreme heat conductions, such as the fast transient laser heating and the ultrahigh heat flux density, the inertia force of thermomass is nonnegligible, causing the inapplicability of the Fourier's law. In this condition the classical entropy production should be modified to the general one. Such modification will influence the expression of entropy, which further changes the definition of temperature based on the second law. If the stagnant temperature is adopted in the expression of entropy, the result is higher than the real one. The reason is that the kinetic energy of thermomass preserves the available energy, which decreases the real entropy of the system. If the static temperature is adopted, the relation of entropy, internal energy, and temperature keeps the same as $1/T = \partial s/\partial e$ even in the condition of non-Fourier heat conduction.

From a microscopic perspective, the derivation based on the phonon Boltzmann equation indicates that the higher order expansion of the phonon state distribution function not only induces the spatial convection term, but also influences the expression of local internal energy. Thus the local energy can be divided into an equilibrium part and a nonequilibrium part. The former corresponds to the local static temperature while the latter is proportional to the kinetic energy of phonon gas, which also induces the distinction between the stagnant and static temperatures.

References

1. Casas-Vazquez J, Jou D (2003) Temperature in non-equilibrium states: a review of open problems and current proposals. Rep Prog Phys 66(11):1937
2. Jou D, Casas-Vázquez J, Lebon G (2010) Extended irreversible thermodynamics. Springer, Berlin
3. Casas-Vázquez J, Jou D (1994) Nonequilibrium temperature versus local-equilibrium temperature. Phys Rev E 49(2):1040
4. Cimmelli VA, Sellitto A, Jou D (2009) Nonlocal effects and second sound in a nonequilibrium steady state. Phys Rev B 79(1):014303
5. Cimmelli VA, Sellitto A, Jou D (2010) Nonequilibrium temperatures, heat waves, and nonlinear heat transport equations. Phys Rev B 81(5):054301

6. Cimmelli VA, Sellitto A, Jou D (2010) Nonlinear evolution and stability of the heat flow in nanosystems: beyond linear phonon hydrodynamics. Phys Rev B 82(18):184302
7. Baranyai A (2000) Temperature of nonequilibrium steady-state systems. Phys Rev E 62 (5):5989
8. Hatano T, Jou D (2003) Measuring nonequilibrium temperature of forced oscillators. Phys Rev E 67(2):026121
9. Clancy LJ (1975) Aerodynamics. Wiley, New York
10. Müller I (1967) On the entropy inequality. Arch Ration Mech Anal 26(2):118–141
11. Sellitto A, Cimmelli VA, Jou D (2013) Entropy flux and anomalous axial heat transport at the nanoscale. Phys Rev B 87(5):054302
12. Guyer RA, Krumhansl JA (1966) Solution of the linearized phonon Boltzmann equation. Phys Rev 148(2):766–778

Chapter 5
Dynamical Analysis of Onsager Reciprocal Relations (ORR)

Abstract The original derivation of the Onsager reciprocal relations requires that the generalized flux in the expression of energy production should be the time derivative of system state variable. However, it was found that the commonly selected fluxes can hardly meet this requirement. In this chapter, the unambiguous definition of the generalized forces and fluxes in the entropy production is presented from the thermomass viewpoint. The linear regression of fluctuation is actually a balance between the inertia force and the friction force. Therefore, the time derivative of state variables is the inertia force rather than the driving force. The state variables are thereby defined as the average displacement of transported quantities during fluctuation. They have the length unit, which is in agreement with the displacement of heat proposed by Onsager. For the coupled transport processes, the reciprocal relations are manifested to be the symmetry of the coefficient matrix between the friction forces and the drift velocities. They can be macroscopically derived through the principles of Galilean invariance and the third law of Newtonian dynamics.

5.1 Basic Assumptions of ORR

In 1931, Onsager [1, 2] derived the reciprocal relations of irreversible processes based on microscopic irreversibility, which became the cornerstone of irreversible thermodynamics. After that, many theoretical and experimental works have proved the reciprocal relations from different perspectives, making it the well-accepted theorem in irreversible thermodynamics. The reciprocal relations indicate that if the generalized forces and fluxes in irreversible processes satisfy the linear relation

$$J_i = \sum_j L_{ij} X_j \tag{5.1}$$

Y. Dong, *Dynamical Analysis of Non-Fourier Heat Conduction and Its Application in Nanosystems*, Springer Theses,
DOI 10.1007/978-3-662-48485-2_5

then the matrix of the phenomenological coefficient is of symmetry

$$L_{ij} = L_{ji} \tag{5.2}$$

The derivation of Onsager Reciprocal Relations (ORR) is inspired by thermoelectric coupling. The anisotropy of thermal conductivity and the three component chemical reactions serve as additional examples. Although the ORR characterizes the linear phenomenological coefficients for irreversible transport processes, its derivation is made in a thermodynamic equilibrium system [1, 2]. The basic assumptions in deriving ORR are (1) microscopic irreversibility and (2) linear regression of fluctuation.

The physical picture described by microscopic irreversibility is that the macroscopic variables such as temperature and heat flux are the functions of microscopic velocities of molecular motion. Quantities such as temperature are the even functions of molecular velocities, which are denoted by α. Quantities like heat flux are the odd functions of molecular velocities, which are denoted by β. Since the molecular motion obeys the dynamic rules, one can assume that at one moment all the molecular velocities are reversed, the α variables will be unchanged while the β variables will change signs (e.g., the heat flux will be reversed). In an equilibrium thermodynamic system, assume that the state of system is Γ_1 and turns to Γ_2 after a period of Δt. Microscopic reversibility requires that at the state of Γ_2, if all the molecular velocities are reversed at one moment, the system should recover the state of Γ_1 after a period of Δt

$$\begin{aligned} \mathcal{H}_{\Delta t}(\Gamma_1) &= \Gamma_2 \\ \mathcal{H}_{-\Delta t}(\Gamma_2) &= \Gamma_1 \end{aligned} \tag{5.3}$$

where $\mathcal{H}$ is the Hamiltonian operator. Equation 5.3 assumes the motion of molecular in the system satisfy the Hamilton dynamics. For two observables, α_1 and α_2, in the system, the microscopic reversibility is expressed as

$$\langle \alpha_1(\Gamma_1)\alpha_2(\Gamma_2)\rangle = \langle \alpha_2(\Gamma_1)\alpha_1(\Gamma_2)\rangle \tag{5.4}$$

The bracket $\langle \ldots \rangle$ denotes the ensemble average. Although microscopic irreversibility arises from the principle of the molecular motion, the fluctuation that randomly happens in the system does not necessarily obey the reversible dynamics. Thus the ensemble average is needed to denote that it is the rule applicable for a large number of events. Microscopic reversibility is a theorem based on the kinetic theory of molecular dynamics. According to the second law of thermodynamics, the macroscopic transports are almost irreversible. Therefore, why the ORR derived from microscopic reversibility is applicable for the macroscopic irreversible transport is still a controversial issue.

The second assumption in the derivation of ORR, the linear regression of fluctuation, is written as

$$\frac{\partial \alpha}{\partial t} = -L\alpha \tag{5.5}$$

where L is the phenomenological coefficient of the linear transport law. One can immediately infer from Eq. 5.5 that the fluctuation α experiences an exponential decay with time. Onsager [1] pointed out that this assumption is in conflict with microscopic irreversibility. With microscopic irreversibility the state variable α is unchanged when all molecular velocities are reversed. Nonetheless, the direction of time evolution will be reversed. One has

$$\frac{\partial \alpha'}{\partial t} = L\alpha' \tag{5.6}$$

which indicates that α increases with time. In real cases the fluctuation cannot keep growing. Therefore, the evolution of α is discontinuous at $t = 0$. The fluctuation emerges at $t = 0$, with an infinite change rate from $t < 0$ to $t > 0$, which does not obey the dynamic process in the system. The sketch of the evolution of α is shown in Fig. 5.1. Note that the term "linear" means that the regression of fluctuation satisfies the linear transport law, namely the decay rate is proportional to α rather than a constant.

It is manifested in Sect. 3.1 that the entropy can be expressed as the production of generalized forces and fluxes. The decomposition of entropy into generalized forces and fluxes in the classical irreversible thermodynamics is to some extent arbitrary. From an intuitive viewpoint the generalized forces are generally selected as the gradient of intensive state variables. The generalized fluxes are defined as the flux of quantities transported. For heat conduction, there is no definition of velocity or force in the classical irreversible thermodynamics. Thus the choice of generalized forces and fluxes are variable. For example, the generalized force and flux in heat conduction can be expressed as [3, 4]

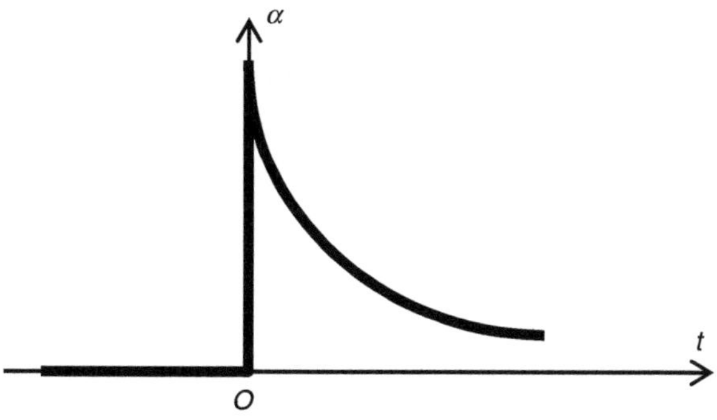

Fig. 5.1 Evolution of a state variable with linear regression of fluctuation

$$X = \nabla\left(\frac{1}{T}\right),\ J = q \tag{5.7a}$$

$$X = -\nabla \ln T,\ J = \frac{q}{T} \tag{5.7b}$$

$$X = -\nabla T,\ J = \frac{q}{T^2} \tag{5.7c}$$

The three formulas in Eqs. 5.7a–5.7c are called the entropy picture, energy picture, and the Fourier picture, respectively. The phenomenological coefficients in these three pictures are thereby different. To impose more restriction on the generalized forces and fluxes, Onsager set the requirement that

$$\boldsymbol{J}_i = \partial\alpha_i/\partial t \tag{5.8a}$$

$$\boldsymbol{X}_i = \partial\Delta S/\partial\alpha_i \tag{5.8b}$$

Thus the generalized flux is expressed by the time derivative of the state variable, α. The generalized force is the derivative of entropy change with respect to α. The entropy change, ΔS, is defined in the equilibrium system, which is the deviation from the equilibrium maximum induced by the fluctuation of α. Hence the time derivative of entropy is expressed as

$$\partial\Delta S/\partial t = \sum_i X_i J_i = \sum_i \sum_j L_{ij} X_i X_j \tag{5.9}$$

The $\partial\Delta S/\partial t$ given by Eq. 5.9 is similar to the entropy production during irreversible processes. Nevertheless, $\partial\Delta S/\partial t$ is defined for the fluctuation in equilibrium systems, while the entropy production, Eq. 3.58, is defined in nonequilibrium systems. Combining Eqs. 5.8a, 5.8b with Eq. 5.9 leads to the proof of ORR for an equilibrium system with fluctuations.

However, for real transport processes, it is found that the ordinary generalized fluxes, such as the heat flux and the momentum flux, can be hardly expressed by the time derivative of state variables. For example, in the heat conduction, the time derivative of the local internal energy (or the fluctuation of temperature) gives the spatial divergence of heat flux, $\nabla\cdot\boldsymbol{q}$. Coleman and Truesdell [5] pointed out that if the generalized flux is not required to be expressed by the time derivative of state variables, one can define

$$J_i' = \sum_j L_{ij} X_j + \sum_j W_{ij} X_j \tag{5.10}$$

where W is a skew-symmetric matrix

$$W_{ij} = -W_{ji}\,(i \neq j),\ W_{ii} = 0 \tag{5.11}$$

Hence the time derivative of entropy does not change

$$\partial \Delta S/\partial t = \sum_i X_i J_i' = \sum_{ij} X_i \left(L_{ij} + W_{ij}\right) X_j = \sum_i X_i J_i \tag{5.12}$$

However, in this case the phenomenological matrix of the coefficient between the generalized forces and fluxes is $L + W$, which is apparently nonsymmetric. The reverse transformation of Eq. 5.10 can also convert a nonsymmetric coefficient matrix, L, to a symmetric matrix. Therefore, one can conclude that the reciprocal relation is only applicable for specified generalized fluxes. The ORR can be broken when the generalized fluxes are not expressed by the time derivative of the state variables [5].

Groot et al. [6] proposed a potential explanation to solve this dilemma. For one-dimensional heat conduction, the time derivative of entropy can be expressed by

$$\frac{\mathrm{d}S}{\mathrm{d}t} = \int_V q \nabla \left(\frac{1}{T}\right) \mathrm{d}V \tag{5.13}$$

The boundaries are adiabatic, $q|_\Gamma = 0$. Assume that the physical properties are constant. Integration by part of Eq. 5.13 gives

$$\frac{\mathrm{d}S}{\mathrm{d}t} = -\int_V \frac{\nabla \cdot q}{T} \mathrm{d}V = \rho \int_V \frac{1}{T} \frac{\partial C_V T}{\partial t} \mathrm{d}V \tag{5.14}$$

The conservation of internal energy of the system requires

$$\int_V \frac{\partial C_V T}{\partial t} \mathrm{d}V = 0 \tag{5.15}$$

In case of small temperature differences, Eq. 5.14 transforms to

$$\frac{\mathrm{d}S}{\mathrm{d}t} = -\frac{\rho}{T_0^2} \int_V (T - T_0) \frac{\partial C_V T}{\partial t} \mathrm{d}V, \tag{5.16}$$

where T_0 is the average temperature. Hence the generalized fluxes and forces can be defined as

$$J = \frac{\partial C_V T}{\partial t}$$
$$X = -\frac{\rho}{T_0^2}(T - T_0) \quad (5.17)$$

The corresponding phenomenological relation is

$$\frac{\partial C_V T}{\partial t} = -\frac{\rho}{T_0^2}\int_V K(T - T_0)\mathrm{d}V, \quad (5.18)$$

where K is the phenomenological coefficient. This derivation is supposed to express the heat flux as the time derivative of state variables. However, Eq. 5.18 is obtained in the adiabatic system. The true heat conduction does not specify the boundary condition and mostly happens in open systems with continuous energy exchange with the environment. Equation 5.18 can be regarded as the attenuation from nonequilibrium to equilibrium in an isolated system with uniform initial temperature profile, which obeys the Fourier's law. The physical picture in the proof of ORR is the attenuation of fluctuation in equilibrium systems, which is a microscopic process and does not obey Fourier's law. In Eq. 5.14 the generalized force is ρ/T. It does not have any direction. The transformation through the constraint of internal energy conservation converts the generalized force into Eq. 5.17. However, the direction of $T - T_0$ is still limited in the thermodynamic phase space. It is not a direction in the Cartesian coordinates.

In the initial derivation of Onsager, the generalized forces for heat conduction and electrical conduction are expressed as $-\nabla T/T$ and $-\nabla\varphi_e$, respectively. The latter is close to the real driving force in the electrical conduction. The temperature gradient is difficult to relate to the real force because $-\nabla T/T$ does not have a unit of force. The essential reason is that there is no mass defined in the heat conduction. If there is some force on the heat, the acceleration would be infinite. Therefore, the generalized force or thermodynamic force is defined in classical irreversible thermodynamics. The irreversible transport cannot be analyzed with the real (Newtonian) force and fluxes.

5.2 Generalized Forces and Fluxes Based on Thermomass Theory

The real forces in heat conduction can be defined through the thermomass theory, namely the driving and friction forces. The driving force is the pressure gradient of thermomass, ∇p_h. The friction force is proportional to the thermomass drift velocity, u_h. They are the force acting on the unit volume, with the unit of N/m^3. The driving force is

$$-\nabla p_{\mathrm{h}} = -\left(\frac{2\gamma_{\mathrm{G}} C_{\mathrm{V}} T}{c^2}\right) e \nabla T / T = -\xi e \nabla T / T, \tag{5.19}$$

where e is the local internal energy density. In Eq. 3.16 it is shown that the coefficient ξ is the square of the ratio of the sound speed over the light speed. If the heat flux is regarded as the generalized flux, the friction force of thermomass, f_{h}, needs to divide ξ to obtain the macroscopic thermomass friction, F_{h}, which still has the unit of volumetric force.

In Sect. 3.2 it is indicated from an energy conversion perspective that ξ is the ratio of thermomass energy over the internal energy, or the ratio of the dissipation rate of thermomass energy over the entropy production rate. From a transport perspective, ξ is the ratio of the real momentum of thermomass over the quasi-momentum, which represents the ratio of the force on the thermomass motion over the force on the conversion from thermal energy to other types of energies. For example, in the thermoelectric effect, the phonon gas and the electron gas exchange both momentum and energy, with conversion between thermal energy and electrical energy. The energy transports from phonons to electrons when the thermal energy converts to electrical energy. In this process there is also momentum balance between the electrons and phonons. The quasi-momentum is conserved in the process of electron–phonon interaction. If one counts the real momentum, it is non-conservative. The missing part is compensated by the real momentum of lattices. The conversion between the thermal and electrical energies will induce the increase in the real momentum of phonons, which is proportional to the increase in the quasi-momentum, with the coefficient, ξ.

The conservation of the quasi-momentum between the phonons and electrons is [7]

$$\hbar \Delta k_{\mathrm{e}} = -\hbar \Delta k_{\mathrm{ph}} + \hbar G, \tag{5.20}$$

where k_{e} is the wave vector of electrons, k_{ph} is the wave vector of phonons, and G is the reciprocal lattice vector. The variables in Eq. 5.20 are in the microscopic scope. Integration in the wave vector space of Eq. 5.20 gives

$$\begin{gathered} \int_{\boldsymbol{k}} \hbar \Delta k_{\mathrm{e}} = -\int_{\boldsymbol{k}} \left(\hbar \Delta k_{\mathrm{ph}} + \hbar G\right) \\ \frac{\Delta \rho_{\mathrm{h}} u_{\mathrm{h}}}{\xi} + \Delta n m_{\mathrm{e}} u_{\mathrm{e}} = p_{\mathrm{Loss}} \end{gathered} \tag{5.21}$$

The quasi-momentum and real momentum of phonons satisfy

$$\int_{\boldsymbol{k}} \hbar k_{\mathrm{ph}} = \frac{1}{v_{\mathrm{s}}^2} \int_{\boldsymbol{k}} \hbar \omega \frac{\partial \omega}{\partial k_{\mathrm{ph}}} = \frac{q}{v_{\mathrm{s}}^2} = \frac{\rho_{\mathrm{h}} u_{\mathrm{h}}}{\xi} \tag{5.22}$$

For electrons they satisfy

$$\int_{k} \hbar k_{\mathrm{e}} = n m_{\mathrm{e}} v_{\mathrm{e}}, \tag{5.23}$$

where n is the charge density, m_{e} is the effective mass of charges, and v_{e} is the drift velocity of electron gas. Equation 5.22 manifests that the quasi-momentum of phonon gas equals the real momentum divided by ξ. Equation 5.23 is the momentum of electron gas, which corresponds to the current density. The integral of G gives p_{Loss}, which is the loss of quasi-momentum. The lost quasi-momentum is absorbed by the lattice, representing the friction of the lattice on the transport of phonon gas. From Eq. 5.21 it is inferred that the friction force on phonon gas and the heat-current interaction force are both $1/\xi$ times larger than the loss of thermomass momentum.

Consider a tank filled with ideal gas. The pressure is

$$p = \frac{1}{3}\rho v_{\mathrm{rms}}^{2}, \tag{5.24}$$

where v_{rms} is the mean square velocity of gas molecular. If the rest mass of molecular is converted into the equivalent energy (electromagnetic irradiation), the pressure on the wall of the tank is

$$p_{\mathrm{R}} = \frac{1}{3}\rho c^{2}, \tag{5.25}$$

where ρc^2 is the total energy of the electromagnetic irradiation. Equation 5.25 is called the pressure of the "electromagnetic cloud" or the "photon gas." It can be inferred that the pressure change induced by the mass-energy conversion is

$$\frac{p_{\mathrm{R}}}{p} = \frac{c^{2}}{v_{\mathrm{rms}}^{2}} \tag{5.26}$$

This ratio is in analogy with $1/\xi$. This derivation manifests that the ξ originates from the mass-energy duality of thermal energy. When the thermal energy exchanges with other forms of energies, the quantity of thermomass varies. Thus the induced force needs to divide ξ to get the macroscopic force. In this section the coupling of heat conduction and other transports is investigated. For convenience, the macroscopic driving force of thermomass is defined as

$$F_{\mathrm{Dh}} = -\frac{e\nabla T}{T} = -\nabla(\rho C_{\mathrm{V}} T) \tag{5.27}$$

and the macroscopic friction force is $\boldsymbol{F}_{\mathrm{h}} = -e\boldsymbol{q}/\kappa T$ (cf. Eq. 3.20).

In Sect. 3.2 it is shown that the entropy production should be the product of $\boldsymbol{F}_{\mathrm{h}}$ and u_{h}, divided by T. It gives the entropy production with a quadratic form of heat flux, which keeps semi-positive definite in non-Fourier heat conduction. The entropy production for other transport processes is also expressed by the product of the real friction force and the corresponding drift velocity, divided by T. Hence the entropy product is written as

$$\sigma^{\mathrm{s}} = -\frac{1}{T}\sum_{i} \boldsymbol{u}_i \cdot \boldsymbol{F}_i \tag{5.28}$$

The generalized flux in classical irreversible thermodynamics can be defined as

$$J_i = \rho_i \boldsymbol{u}_i \tag{5.29}$$

For heat conduction, electrical conduction, and mass component diffusion, ρ_i represents the internal energy density, e, with the unit J/m^3, the charge density ρ_{e}, with the unit C/m^3, the component density ρ_{k}, with the unit kg/m^3. The unit of the drift velocity is m/s.

The generalized forces and fluxes satisfy the linear phenomenological law in classical irreversible thermodynamics. By defining the real forces and velocities in transport processes, this linear relation can be rewritten as

$$\boldsymbol{u}_i = -\sum_{j} \Lambda_{ij} \cdot \boldsymbol{F}_j \tag{5.30}$$

This indicates the linear relation between the friction forces and the drift velocities. Λ is the matrix of friction coefficients. The minus in Eq. 5.30 shows that the friction forces are in the opposite directions of the drift velocities. Compared with Eq. 5.1, Λ has the form

$$\Lambda_{ij} = \frac{L_{ij}}{T\rho_i\rho_j} \tag{5.31}$$

Thus the symmetry of Λ is the same as L. The unit of Λ_{ij} is m^3·s/kg. Note that the L is especially defined based on generalized fluxes in Eq. 5.29 and their corresponding generalized forces. If one adopts other types of decomposition of generalized forces and fluxes, for instance, putting the coefficient T^{-1} in generalized fluxes, the unit of L and its relation to Λ will be changed. However, the unit of Λ keeps the same since the definition of forces is unambiguous.

The proof of ORR requires that the generalized fluxes are the time derivative of state variables. In the following the expression of state variables, α, are explored. From Eqs. 5.28 to 5.30, one infers that the generalized fluxes are related to the drift velocities and the drift velocities are linear to the friction forces. Therefore, the time derivative of α can be expressed as the linear combination of friction forces

$$\frac{\partial \alpha_i}{\partial t} = -\sum_j \Lambda_{ij} \cdot \boldsymbol{F}_j \tag{5.32}$$

Onsager [8] required that α has the following features: it is a macroscopic observable defined in a subsystem containing a large amount of molecular; it is the algebra summation of microscopic molecular variables, hence its fluctuation obeys the Gaussian distribution; it is the even function of molecular velocities. When the molecular velocities are suddenly reversed, it keeps unchanged. Therefore, Eq. 5.32 actually represents the linear regression of fluctuations. Since F_j is the friction force on the flow of linear transports, this equation indicates that α is attenuated by the friction force. Recall that F_j is the function of the generalized fluxes, which gives

$$\frac{\partial \alpha_i}{\partial t} = u_i = \frac{J_i}{\rho_i} \tag{5.33}$$

On the other hand, consider a situation where the driving forces in the steady linear transports, for instance the temperature gradient, are removed suddenly, the fluxes in system would experience a decay process. In terms of the drift velocities, this process is described as

$$R_i \rho_i \frac{\partial \boldsymbol{u}_i}{\partial t} = \boldsymbol{F}_i = -\sum_j \left(\Lambda^{-1}\right)_{ij} \boldsymbol{u}_j, \tag{5.34}$$

where $R_i\rho_i$ depicts the ability of resistance against the forced acceleration, i.e., inertia

$$R_i = 1/v_s^2, \text{ (heat conduction)} \tag{5.35a}$$

$$R_i = m_e/\varepsilon, \text{ (electrical conduction)} \tag{5.35b}$$

$$R_i = 1, \text{ (mass component conduction)}, \tag{5.35c}$$

where ε is the charge of carriers. The characteristic relaxation time of u_i can be extracted from Eq. 5.34

$$\tau_{ij} = \Lambda_{ij} R_j \rho_j \tag{5.36}$$

This relaxation behavior can be understood as when $i = j$, the transport flux decays under the effect of the friction proportional to its drift velocity; when $i \neq j$, the transport flux induces other types of fluxes. These fluxes are also impeded by frictions, which cause additional friction force on the ith transport flux. The latter friction can be regarded as the friction induced by the relative motion of the ith and jth transport fluxes. In the classical Fourier's law, the heat flux is determined by the temperature gradient. So the heat flux is zero when the temperature gradient is suddenly removed. The thermal energy has inertia according to the thermomass theory. Therefore, with the temperature gradient suddenly removed, the heat flux will experience a free decay process. Onsager [1] pointed out for Eq. 5.5 that the linear transport law is only an approximation of heat conduction which neglects the time needed for the acceleration of heat flux. Hence, the picture of Eq. 5.34 is the free decay of transport flux due to friction without the driving forces. The friction forces are linear to the drift velocities in this procedure. The thermomass theory claims that the relation between the friction forces and velocities does not depend on the driving force or the inertia force. In other words, such relation in the free decay of heat flux is the same as that in the steady conduction. This picture is identical to Onsager's assumption of the linear regression of fluctuation, Eq. 5.5.

Note that the linear regression of fluctuation is assumed in the case of small fluctuations. The remaining kinetic energy only attenuates without inducing other effects. With large fluctuations, the evolution may contain the oscillation, i.e., the remaining kinetic energy is not completely dissipated by the friction but partly converted into potential energy. Marconi et al. [9] derived through the Lagrange variational principle that the white noise satisfies the Gaussian distribution in case of small fluctuations, with Eq. 5.5 as the most probable path from the initial α_0 to α_t after a time period of t. For large fluctuations the decay of fluctuation can be described by a hydrodynamic equation

$$\frac{\partial \alpha}{\partial t} = -L\alpha + A, \tag{5.37}$$

where A is the orthogonal to the thermodynamic forces. Due to the orthogonality, A does not contribute to the entropy production. The time reversibility is broken by A. Gabrielli et al. [10] derives the ORR based on the deterministic equation in the Hamiltonian system. The time evolution equation of fluctuation contains a term in addition to the attenuation caused by friction. This additional term does not obey the microscopic reversibility assumed by Onsager. From the thermomass viewpoint, the driving force and inertia force have no contribution to the entropy production. They can be regarded as orthogonal to the friction force. The partial differential operator, $\partial/\partial t$, can be substituted by the material differential, D/Dt, to include the spatial inertia effect. Then A in Eq. 5.37 plays a similar role to the driving force, ∇T. In far from equilibrium or steady nonequilibrium systems, the instantaneous or the steady driving force can break the time symmetry. This effect is equivalent to shifting the thermodynamic equilibrium so that the spontaneous fluctuation couples with the force field existing in the system. However, as long as the coupling is unrelated to friction, it will not contribute to the entropy production.

Comparison of Eq. 5.32 with Eq. 5.34 can extract the expression of α_i

$$\alpha_i = -\sum_i \tau_{ij} u_j \tag{5.38}$$

Thus the state variables can be defined by the product of drift velocities and the relaxation times. The time derivative of α_i is the drift velocity u_i. In Onsager's derivation, the time derivative of state variables gives the generalized fluxes. For the usually chosen generalized fluxes, e.g., heat flux, electrical flux, and mass component flux, the relation between α_i defined in Eq. 5.38 and J_i is

$$\frac{\partial \rho_i \alpha_i}{\partial t} = J_i, \tag{5.39}$$

where $\rho_i \alpha_i$ also satisfies the requirement of Onsager on state variables [8]. Nevertheless, the $\alpha_i s$ in Eq. 5.38 have the same unit, i.e., meter. If $\rho_i \alpha_i s$ are chosen as the state variables, they will have different units. Since the ensemble average of the fluctuation of state variables satisfies [1]

$$\left\langle \alpha_i^2 \right\rangle = \left\langle \alpha_j^2 \right\rangle \tag{5.40}$$

Using $\alpha_i s$ with the same unit is more reasonable and convenient.

In heat conduction α_i can be understood as the displacement of heat. In Onsager's derivation [1], α_i is also defined as the displacement of the weight center of heat. Because there is no concept of mass in heat conduction, the following research inclined to define α_i as the local temperature fluctuation. Astumian [11, 12] defined the average distance travelled by a particle during a period of fluctuation as the state variable in studying the ORR in the Brownian molecular sieve. The local internal energy is the even function of molecular velocities. The random motion of molecular will cause the random displacement of the energy, which can be regarded as the displacement of heat, α_i. Similarly, the charge displacement and mass component displacement can be defined in the electrical conduction and mass diffusion processes. Such displacement causes the nonuniform distribution of physical quantities.

In Eqs. 5.16–5.18, the generalized fluxes are expressed by the time derivative of $C_V T$ in one-dimensional adiabatic system [6]. The corresponding state variable can be compared with α_i defined in Eq. 5.38. Consider the heat conduction in one-dimensional uniform system with length $2l$. Assume that at the initial moment the temperature distribution satisfies

$$T(x) = T_0 + a_1 \cos\left(\frac{\pi x}{l}\right) \tag{5.41}$$

If the heat conduction obeys the Fourier's heat conduction law and the two ends of the system are adiabatic, then the heat flux at this moment is

$$q(x) = -\kappa T(x) = \kappa a_1 \frac{\pi}{l} \sin\left(\frac{\pi x}{l}\right) \tag{5.42}$$

The rate of temperature evolution is

$$\frac{\partial \rho C_V T}{\partial t} = -\nabla q(x) = -\kappa a_1 \left(\frac{\pi}{l}\right)^2 \cos\left(\frac{\pi x}{l}\right) \tag{5.43}$$

Since

$$a_1 = \frac{T(x) - T_0}{\cos\left(\frac{\pi x}{l}\right)} \tag{5.44}$$

Equation 5.43 can be written as

$$\frac{\partial \rho C_V T}{\partial t} = -\kappa \left(\frac{\pi}{l}\right)^2 (T - T_0) \tag{5.45}$$

Then the temperature evolution is

$$\frac{T(x,t) - T_0}{T(x,0) - T_0} = \exp\left(-\frac{\kappa}{\rho C_V} \frac{\pi^2}{l^2} t\right) \tag{5.46}$$

The temperature difference, $T - T_0$, can be regarded as the temperature disturbance in the system, thus the characteristic relaxation time is

$$\tau^* = \frac{\rho C_V}{\kappa} \frac{l^2}{\pi^2} \tag{5.47}$$

On the other hand, the thermomass drift velocity at the initial moment is

$$\begin{aligned} u_h(x,0) &= \kappa a_1 \frac{\pi}{l} \sin\left(\frac{\pi x}{l}\right) \frac{1}{\rho C_V T_0} \\ &= \kappa \frac{\pi}{l} \tan\left(\frac{\pi x}{l}\right) \frac{T(x,0) - T_0}{\rho C_V T_0} \end{aligned} \tag{5.48}$$

In analogy to Eq. 5.38, the displacement of heat, α^*, is defined based on Eqs. 5.47 and 5.48

$$\begin{aligned} \alpha^*(x,0) &= -\frac{l}{\pi} \frac{T(x,0) - T_0}{T_0} \tan\left(\frac{\pi x}{l}\right) \\ &= -\frac{l}{\pi} \frac{a_1}{T_0} \sin\left(\frac{\pi x}{l}\right) = \left(\frac{l}{\pi}\right)^2 \frac{\nabla T(x,0)}{T_0} \end{aligned} \tag{5.49}$$

This depends on the temperature profile in the system. Therefore, the temperature distribution seems to be the state variable corresponding to the heat flux. However, the derivation of α^* is based on the Fourier heat conduction law, which is a different physical picture from Eq. 5.34.

Comparison of Eq. 5.45 with Eq. 5.17 shows that the generalized forces and fluxes defined by Groot et al. [6] satisfy

$$J = \kappa \frac{T_0^2}{\rho^2}\left(\frac{\pi}{l}\right)^2 X \tag{5.50}$$

For more general cases, the temperature distribution can be written in Fourier series

$$T(x,0) = T_0 + \sum_n a_n \cos\left(\frac{n\pi x}{l}\right) + \sum_n b_n \sin\left(\frac{n\pi x}{l}\right) \tag{5.51}$$

Thus Eq. 5.45 can be reformed as

$$\begin{aligned} \frac{\partial \rho C_V T}{\partial t} &= -\kappa\left[\sum_n a_n \left(\frac{n\pi}{l}\right)^2 \cos\left(\frac{n\pi x}{l}\right) + \sum_n b_n \left(\frac{n\pi}{l}\right)^2 \sin\left(\frac{n\pi x}{l}\right)\right] \\ \frac{\partial T}{\partial t} &= -\frac{1}{\tau^*}\left[\sum_n n^2 a_n \cos\left(\frac{n\pi x}{l}\right) + \sum_n n^2 b_n \sin\left(\frac{n\pi x}{l}\right)\right] \end{aligned}, \tag{5.52}$$

where τ^* is defined as Eq. 5.47. In this condition, the temperature evolution cannot be expressed by a single relaxation time. With the increase of the characteristic wave vector (proportional to n/l), the corresponding relaxation time decreases (inversely proportional to n^2). Each component relaxes individually with the corresponding relaxation time. Therefore, Eq. 5.49 can be reformed as

$$\alpha^*(x,0) = -\sum_n \frac{l}{n\pi}\frac{a_n}{T_0}\sin\left(\frac{n\pi x}{l}\right) - \sum_n \frac{l}{n\pi}\frac{b_n}{T_0}\cos\left(\frac{n\pi x}{l}\right) \tag{5.53}$$

This is determined by the initial temperature profile, with a_n, b_n obtained from the Fourier transformation. Meanwhile Eq. 5.50 turns to

$$J = \kappa \frac{T_0^2}{\rho^2} k^2 X, \tag{5.54}$$

where k is the average characteristic wave vector of the initial temperature profile. The relation between forces and fluxes, i.e., Eqs. 5.50 and 5.54, are linear. However, the proportional coefficient depends not only on the physical properties (e.g., thermal conductivity), but also on k. In other words, the phenomenological coefficient based on Groot et al.'s [6] derivation depends on the boundary condition, which cannot be regarded as a physical property of medium.

From the thermomass viewpoint, the attenuation of fluctuation is the balance between the inertia and friction force. This process does not obey the Fourier's heat conduction law. The temperature profile depicted in Eq. 5.41 is macroscopic. It is much larger than the microscopic fluctuation and is not suited for the fluctuation–dissipation theory. Consider an equilibrium system with the fluctuational flux, δq. The displacement of heat, α, can be derived from Eqs. 5.34 to 5.38

$$\alpha = -\Lambda_{qq}\frac{\delta q}{v_s^2} \tag{5.55}$$

Inserting the propagation speed in the thermomass, Eq. 2.21, one has

$$\alpha = -\tau_{\mathrm{TM}}\Lambda_{qq}\frac{\rho C_{\mathrm{V}}}{\kappa}\delta q, \tag{5.56}$$

where the fiction coefficient Λ_{qq} can be obtained from Eq. 3.20

$$\Lambda_{qq} = -\frac{u_{\mathrm{h}}}{F_{\mathrm{h}}} = \frac{\kappa}{(\rho C_{\mathrm{V}})^2 T} \tag{5.57}$$

Substitution of Eq. 5.57 into Eq. 5.56 gives

$$\alpha = -\tau_{\mathrm{TM}}\frac{\delta q}{\rho C_{\mathrm{V}} T} = -\tau_{\mathrm{TM}}\delta u_{\mathrm{h}} = -\frac{\kappa}{\rho C_{\mathrm{V}}}\frac{1}{2\gamma_{\mathrm{G}} C_{\mathrm{V}} T}\delta u_{\mathrm{h}} \tag{5.58}$$

Compared with Eq. 5.49, it can be seen that the difference comes from the relaxation time. τ^* is inversely proportional to the thermal diffusivity, $\kappa/\rho C_{\mathrm{V}}$, in Eq. 5.47. In contrast, τ_{TM} is proportional to the thermal diffusivity in Eq. 5.56. The former indicates that the nonuniform distribution of temperature relaxes to the uniform one in a faster speed with a larger thermal diffusivity. The latter implies that the attenuation of fluctuational heat flux will be slower with a larger thermal diffusivity. The former is macroscopic heat conduction while the latter is microscopic fluctuation. In general, τ^* is much larger than τ_{TM}. For the silicon at room temperature, if the characteristic wavelength of temperature profile reduces to the magnitude of 100 nm, the two relaxation times will be of the same order, 10^{-10} s. However, in such small scale the transport would be ballistic-diffusive and the heat conduction could deviate from Fourier's law.

From the above analysis one infers that the displacement of heat is generally small. It corresponds to the amplitude of the spontaneous fluctuation of heat in equilibrium systems. Therefore, it cannot be expressed by macroscopic variables, e.g., Eq. 5.49.

The generalized flux based on the definition of Eq. 5.58 is

$$\boldsymbol{J} = \frac{\partial e\alpha}{\partial t} = e\delta \boldsymbol{u}_{\mathrm{h}}, \tag{5.59}$$

while the generalized force is

$$\boldsymbol{F} = -\left(\Lambda^{-1}\right)_{qq}\delta\boldsymbol{u}_{\mathrm{h}} = -\frac{(\rho C_{\mathrm{V}})^{2}T}{\kappa}\delta\boldsymbol{u}_{\mathrm{h}} \tag{5.60}$$

The generalized force and flux satisfy

$$\boldsymbol{J} = -\frac{\kappa}{\rho C_{\mathrm{V}}}\boldsymbol{F} \tag{5.61}$$

The corresponding phenomenological coefficient is the negative thermal diffusivity. It is the physical property of materials and does not depend on the initial temperature profile. In ORR, the coefficient matrix is positive definite. Here the coefficient is negative. The reason is that the generalized force in ORR is the driving force while in Eq. 5.61 it is the friction force. The friction force is in the opposite direction of driving force, thereby the coefficient turns to negative.

Based on the definition by Eqs. 5.33 and 5.34, the entropy change due to fluctuation is

$$\Delta S = -\frac{1}{2}\sum_{i,j}\frac{R_{i}\rho_{i}u_{i}u_{j}}{T} \tag{5.62}$$

Equation 4.34 shows that the expression of entropy should be modified in the presence of heat flux in nonequilibrium transport. The distinction is

$$\Delta S = S - S_{0} = -\frac{1}{2}\frac{\tau_{\mathrm{TM}}q^{2}}{\kappa T^{2}} \tag{5.63}$$

Inserting Eqs. 5.34 and 5.57 into Eq. 5.62 gives

$$\Delta S = -\frac{1}{2}\frac{\tau_{\mathrm{TM}}u_{\mathrm{h}}^{2}}{\Lambda_{qq}T} = -\frac{1}{2}\frac{\tau_{\mathrm{TM}}u_{\mathrm{h}}^{2}}{T}\frac{(\rho C_{\mathrm{V}})^{2}T}{\kappa} = -\frac{1}{2}\frac{\tau_{\mathrm{TM}}q^{2}}{\kappa T^{2}} \tag{5.64}$$

which is identical to Eq. 5.63. As Eq. 5.64 describes the heat flux fluctuation in equilibrium systems, the balance between the inertia force and friction force is

$$\tau_{\mathrm{TM}}\frac{\partial q}{\partial t} + q = 0 \tag{5.65}$$

Therefore, the entropy production is

$$\sigma^{s} = \frac{\partial \Delta S}{\partial t} = -\frac{\tau_{\mathrm{TM}}q}{\kappa T^{2}}\frac{\partial q}{\partial t} = \frac{\tau_{\mathrm{TM}}q^{2}}{\kappa T^{2}} \tag{5.66}$$

which is in consistent with the analysis based on thermomass theory, Eq. 4.34. Furthermore, the entropy production of coupled multiple transports can be obtained as

$$\sigma^s = \frac{\partial \Delta S}{\partial t} = -\frac{1}{T}\sum_{i,j} u_j\left(\Lambda^{-1}\right)_{ij}\frac{\partial \tau_{ij} u_i}{\partial t} = \frac{1}{T}\sum_{i,j}\left(\Lambda^{-1}\right)_{ij} u_i u_j \tag{5.67}$$

Thus the entropy production is the quadratic form of drift velocities of transports. As long as Λ is positive definite, the entropy production keeps nonnegative for all kinds of transports.

5.3 Macroscopic Proof of ORR

From Eq. 5.30 it can be seen that the reciprocal relations are in essence the symmetry of the coefficient matrix, Λ, which relates the friction forces and drift velocities. This coefficient matrix does not depend on the initial state or the evolution rate of transport processes. The off-diagonal terms in Λ represent the cross effect of various transport processes. According to Eq. 5.21 the interaction between the phonon gas and electron gas satisfies the momentum conservation. At the same time, the phonon gas and electron gas are both impeded by other scatterings in materials, which absorb their momentum. Thus it is inferred that the cross effect among various transport processes is the momentum exchange process induced by mutual friction. For example, the temperature gradient drives heat flux. If the motion of thermomass transfers some momentum to the charge carriers, the heat flux then drives the motion of charges, which thereafter induces electrical current. The overall effect of this process is the temperature gradient drives both heat flux and electrical current. The behavior is described as: one generalized force causes multiple generalized fluxes; one generalized flux is the result of multiple generalized forces. Nevertheless, by defining the generalized forces as the friction forces in transport processes, this behavior can be re-narrated as: one flow is impeded by multiple frictions; these frictions can serve as the driving force of other transports. Therefore, the friction on the ith transport flux can be written as

$$-\boldsymbol{F}_i = \left(\Lambda^{-1}\right)_{ii}(\boldsymbol{u}_i - \boldsymbol{u}_0) + \sum_{j\neq i}\left(\Lambda^{-1}\right)_{ij}(\boldsymbol{u}_i - \boldsymbol{u}_j), \tag{5.68}$$

where u_0 is the moving speed of the medium, and u_i and u_j are the drift velocity of the corresponding transport fluxes. The first term on the right-hand side of Eq. 5.68 is the interaction between the transport flux and the medium. Such interaction rises from the reduction of phonon gas momentum due to R scatterings. The second term on the right-hand side of Eq. 5.68 is the interaction between the ith transport flux and other fluxes, which is proportional to the relative velocity, $\boldsymbol{u}_i - \boldsymbol{u}_j$. The action equals the reaction principle, i.e., the force on ith fluid acted by jth fluid equals the inverse force on jth fluid acted by ith fluid; one obtains

$$\left(\Lambda^{-1}\right)_{ij}(\boldsymbol{u}_i - \boldsymbol{u}_j) = -\left(\Lambda^{-1}\right)_{ji}(\boldsymbol{u}_j - \boldsymbol{u}_i) \tag{5.69}$$

which immediately gives the reciprocal relation

$$\left(\Lambda^{-1}\right)_{ij} = \left(\Lambda^{-1}\right)_{ji} \tag{5.70}$$

Equation 5.68 also satisfies the Galilean invariance principle, i.e., the friction force is the function of relative velocities among various fluxes. Hence, the friction force is invariable in all the inertia references. The drift velocities discussed here are all much less than the light speed, so the Galilean invariance is applicable. Therefore, the derivation of ORR is based on two principles, namely the Galilean invariance and the third law of Newtonian dynamics—action equals reaction, and one assumption that the interactions between the fluxes are proportional to their relative velocities. Compared with Onsager's original derivation, only the macroscopic principles are used in this proof without any assumption of the microscopic fluctuations. In fact, it is a controversial issue whether the microscopic fluctuation satisfies the linear regression or the microscopic irreversibility. Consider that reciprocal relations are mostly proved in macroscopic irreversible transport processes, a macroscopic derivation of ORR can avoid the controversy in the microscopic scope.

In physics, the third law of Newtonian dynamics is another face of the momentum conservation. Actually, the momentum conservation can be related to the microscopic irreversibility. Onsager [1] presented the microscopic reversibility as

$$\langle \alpha_1(t)\alpha_2(t+\Delta t)\rangle = \langle \alpha_2(t)\alpha_1(t+\Delta t)\rangle, \tag{5.71}$$

where the time evolution of state variables, α_1 nad α_2 (e.g. displacement of heat), in a period of Δt can be written in a first order expansion form

$$\alpha_2(t+\Delta t) = \alpha_2(t) + \frac{\partial \alpha_2}{\partial t}\Delta t \tag{5.72}$$

Inserting it into Eq. 5.71 gives

$$\langle \alpha_1(t)\alpha_2(t+\Delta t)\rangle = \left\langle \alpha_1(t)\frac{\partial \alpha_2}{\partial t}(t)\right\rangle \Delta t = -L_{21}\langle \alpha_1^2\rangle \Delta t \tag{5.73}$$

This transformation uses the linear regression of fluctuation, namely

$$\frac{\partial \alpha_2}{\partial t} = -L_{21}\alpha_1 - L_{22}\alpha_2 \tag{5.74}$$

Similarly,

$$\langle \alpha_2(t)\alpha_1(t+\Delta t)\rangle = \left\langle \alpha_1(t)\frac{\partial \alpha_2}{\partial t}(t)\right\rangle \Delta t = -L_{12}\langle \alpha_2^2 \rangle \Delta t \quad (5.75)$$

Based on Eqs. 5.40, 5.71, and 5.75 it can be derived that

$$L_{12} = L_{21} \quad (5.76)$$

Thus it is the proof based on the microscopic reversibility. Note that Eq. 5.71 can be reformed as

$$\langle \alpha_1(t)[\alpha_2(t) + \Delta\alpha_2(\Delta t)]\rangle = \langle \alpha_2(t)[\alpha_1(t) + \Delta\alpha_1(\Delta t)]\rangle \quad (5.77)$$

According to the principle of action equals reaction, the increase of α_2 during Δt should equal the decrease of α_1 during the same period. Hence

$$\langle \alpha_1(t)\Delta\alpha_2(\Delta t)\rangle = -\langle \alpha_1(t)\Delta\alpha_1(\Delta t)\rangle \quad (5.78)$$

Thereafter, the derivation goes as

$$\begin{aligned}
&\langle \alpha_1(t)[\alpha_2(t) + \Delta\alpha_2(\Delta t)]\rangle \\
&\quad = \langle \alpha_1(t)\alpha_2(t) - \alpha_1(t)\Delta\alpha_1(\Delta t)\rangle \\
&\quad = \langle \alpha_1(t)\alpha_2(t) - \alpha_2(t)\Delta\alpha_2(\Delta t)\rangle \\
&\quad = \langle \alpha_2(t)[\alpha_1(t) + \Delta\alpha_1(\Delta t)]\rangle
\end{aligned} \quad (5.79)$$

Note that here the relation used is

$$\langle \alpha_1(t)\Delta\alpha_1(\Delta t)\rangle = \langle \alpha_2(t)\Delta\alpha_2(\Delta t)\rangle, \quad (5.80)$$

which is equivalent to Eq. 5.40. Equation 5.79 manifests that the microscopic reversibility actually can be also derived from the momentum conservation principle. Because Onsager elucidates α as observables with enough number of molecular, it is inferred that the microscopic reversibility is not completely "microscopic" but has some macroscopic features.

Here the interaction between two fluxes is assumed to be proportional to the relative velocity, which is the simplest linear assumption. In more general cases, the interaction between two fluxes can be expanded into the polynomial of the relative velocities. The action equals reaction principle is interpreted as

$$\sum_{n=1}^{\infty} (\Lambda^{-1})_{ij}^{(n)} (\boldsymbol{u}_i - \boldsymbol{u}_j)^n = -\sum_{n=1}^{\infty} (\Lambda^{-1})_{ji}^{(n)} (\boldsymbol{u}_j - \boldsymbol{u}_i)^n \quad (5.81)$$

Since

$$\boldsymbol{u}_i - \boldsymbol{u}_j,\ (\boldsymbol{u}_i - \boldsymbol{u}_j)^2,\ \ldots,\ (\boldsymbol{u}_i - \boldsymbol{u}_j)^n \tag{5.82}$$

are linear independent, the coefficients of different orders in Eq. 5.81 should equal respectively, i.e.,

$$\left(\Lambda^{-1}\right)_{ij}^{(n)} = \left(\Lambda^{-1}\right)_{ij}^{(n)} \tag{5.83}$$

This relation can be used to analyze the higher order cross effects beyond the linear transports, such as the molecular sieve and Brownian motor. Note that Eqs. 5.68 and 5.81 do not require $\boldsymbol{F}_{\mathrm{i}}$, $\boldsymbol{u}_{\mathrm{i}}$, $\boldsymbol{u}_{\mathrm{j}}$ are in the same direction, thus the forces can couple with fluxes in different directions, which could be the case of the asymmetry thermal conductivity or the molecular sieves.

The irreversible transport processes discussed here are mainly heat conduction, electrical conduction, and the mass component diffusion. The couplings among these three processes are the most observable and applicable. Other transport processes, such as the momentum transport and the chemical reaction, are different from those discussed above. According to the Curie's principle, the momentum transport and the chemical reaction do not couple with the vectorial transport processes since they are of different orders. For tensorial transport such as the momentum diffusion, its drift velocity is hard to define. In momentum transport the dissipation comes from viscosity. The viscous stress is proportional to the velocity gradient in the fluid. Therefore, the friction in the simple shear flow is

$$\begin{aligned} -\boldsymbol{F}(y) = {} & \left(\Lambda^{-1}\right)_{y,y+\Delta y}(\boldsymbol{u}_{\mathrm{f}}(y) - \boldsymbol{u}_{\mathrm{f}}(y+\Delta y)) \\ & + \left(\Lambda^{-1}\right)_{y,y-\Delta y}(\boldsymbol{u}_{\mathrm{f}}(y) - \boldsymbol{u}_{\mathrm{f}}(y-\Delta y)) \end{aligned} \tag{5.84}$$

If Λ^{-1} represents the viscosity coefficient, Eq. 5.84 is Newton's viscous law. For ordinary uniform fluid, the viscosity is constant. One does not need the reciprocal relation. However, for the asymmetry fluids such as the polymer solution or liquid crystal, the viscosity is not constant but could be a matrix. In this case the reciprocity analysis can be useful.

5.4 Conclusion

The traditional derivation of ORR has some limitations. First, the generalized flux cannot be expressed by the time derivative of state variables, which induces the potential breakdown of ORR. Second, the definitions of generalized forces and fluxes are to some extent arbitrary. Third, the microscopic reversibility and the linear regression of fluctuation are the assumptions for microscopic fluctuations.

The ORR describes the macroscopic transport coefficients. The consistency between the microscopic and macroscopic processes is controversial.

The thermomass theory reveals that the ORR can be understood as the symmetry of the coefficient matrix between the friction forces and transport fluxes. Then the generalized forces are the friction forces, which are real Newtonian forces. The generalized fluxes can be decomposed into the density and the drift velocity. It has shown that the linear regression of fluctuation is the balance between the inertia force and friction force in equilibrium systems. The time derivative of the generalized fluxes represents the inertia forces. The state variables of systems can be defined as the length quantity such as the "displacement of heat." These quantities are the characteristic displacement or the length of random walk of heat (thermomass) during fluctuation. Thus the dilemma that the generalized fluxes cannot be expressed as the time derivative of state variables is solved.

The Fourier heat conduction is the balance between the driving force and friction force of thermomass. The balance between the inertia force and the friction is a non-Fourier heat conduction process. Connection of the linear regression of the microscopic fluctuation to the macroscopic transport coefficients implies that the friction-velocity relation is unchanged in Fourier and non-Fourier heat conduction processes, which agrees with the thermomass theory. Therefore, the state variables corresponding to the heat flux can only be found through the non-Fourier heat conduction processes.

Based on the picture of thermomass theory, the ORR can be macroscopically derived. The Galilean invariance and the third law of Newtonian dynamics—action equals reaction are the basic principles used in the proof. The friction between various fluxes is assumed to be proportional to their relative velocity. This proof does not require any assumption on the microscopic features of fluctuations.

In Onsager's proof in 1931 [1, 2], it was at the stage of equilibrium thermodynamics. The irreversible transports that actually happen in nonequilibrium systems were investigated with the assumptions in equilibrium systems, e.g., microscopic reversibility and linear regression of fluctuation. It could be controversial whether the rule in equilibrium system can be used to characterize the nonequilibrium one. In contrast, the proof based on the dynamic analysis starts from the nonequilibrium system, which avoids this controversy.

References

1. Onsager L (1931) Reciprocal relations in irreversible processes, I. Phys Rev 37(4):405
2. Onsager L (1931) Reciprocal relations in irreversible processes, II. Phys Rev 38(12):2265
3. Gyarmati I, Gyarmati E (1970) Nonequilibrium thermodynamics. Springer, Berlin
4. Zeng DL (1991) Engineering nonequilibrium thermodynamics. Sci Press, Beijing
5. Coleman BD, Truesdell C (1960) On the reciprocal relations of Onsager. J Chem Phys 33 (1):28–31
6. Mazur P, de Groot SR (1963) Nonequilibrium thermodynamics. Amsterdam
7. Grimvall G (1981) The electron phonon interaction in metals. Amsterdam

8. Onsager L, Machlup S (1953) Fluctuations and irreversible processes. Phys Rev 91(6): 1505–1512
9. Marconi UMB, Puglisi A, Rondoni L et al (2008) Fluctuation–dissipation: response theory in statistical physics. Phys Rep 461(4):111–195
10. Gabrielli D, Jona. Lasinio G, Landim C (1996) Onsager reciprocity relations without microscopic reversibility. Phys Rev Lett 77(7):1202–1205
11. Astumian RD (2008) Reciprocal relations for nonlinear coupled transport. Phys Rev Lett 101 (4):046802
12. Astumian RD (2009) Generalized fluctuation.dissipation and reciprocal relations for Brownian sieves and molecular machines. Phys Rev E 79(2):02111

Chapter 6
Dynamical Analysis of Heat Conduction in Nanosystems and Its Application

Abstract The heat conduction in nanosystems deviates from the Fourier's law. The effective thermal conductivity depends on the system size, which is understood as a result of the phonon-boundary scattering. It is proposed in this chapter that the phonon gas (the form of thermomass in dielectric materials) exhibits the viscosity and rarefication effects in nanosystems. The viscosity effect induces the nonuniform heat flux profile at each cross section, which reflects the extra boundary resistance on heat conduction. On the other hand, in nanosystems, the boundary scattering shortens the MFPs of normal processes, which reduces the effective viscosity of phonon gas. With the modification of effective MFPs, the prediction models for the in-plane heat conduction in Si nanofilms and nanowires are developed. The numerical prediction agrees well with the experiments. For the cross-plane heat conduction, the phonon distribution function transits from the ballistic one at the boundary to the diffusive one away from the boundary. A ballistic–diffusive model is thereby established based on the phonon Boltzmann equation. This model agrees well with the molecular dynamics simulations for the cross-plane thermal conductivity (CPTC) of Si nanofilms.

The heat conduction in nanosystems is different from that in bulk materials. Such unique behavior has been studied by many researchers in the past decade. It is commonly agreed that the heat conduction in nanosystems does not obey the Fourier's law. Nevertheless, the effective thermal conductivity defined based on the Fourier's law is still used to describe the experimental data, which is

$$\kappa_{\mathrm{eff}} = \frac{Ql}{A(T_{\mathrm{h}} - T_{\mathrm{c}})}, \tag{6.1}$$

where Q is the total heat flux passing through the cross section, $T_{\mathrm{h}} - T_{\mathrm{c}}$ is the temperature difference across the system, l is the distance between the hot side and the cold side, and A is the area of the cross section. The experiments show that the low-dimensional materials such as the graphene, and CNT have much higher thermal conductivity than bulk materials, which increases with the system length. On the other hand, the semiconductor nanosystems such as the silicon nanowires and nanofilms exhibit much lower thermal conductivity compared with the bulk

Y. Dong, *Dynamical Analysis of Non-Fourier Heat Conduction and Its Application in Nanosystems*, Springer Theses,
DOI 10.1007/978-3-662-48485-2_6

counterparts, which decreases with the characteristic size of the system. The size dependence of the effective thermal conductivity is called the size effect of nanosystems. In this chapter, the motion of phonon gas in nanosystems will be analyzed with the thermomass theory. Thereafter the models predicting the effective thermal conductivity of nanosystems will be developed.

6.1 Existing Models for Heat Conduction in Nanosystems

The thermal resistance is induced by the phonon scattering in dielectric materials. According to Debye's expression, Eq. 1.12, the thermal conductivity of material is proportional to the MFP of *R* processes, λ_R. The *R* processes include the phonon–phonon Umklapp scattering and impurity scattering, which are independent of the material size. In the uniform homogeneous medium, the average MFP of phonon scattering can be regarded as a constant. In this case the heat conduction obeys the Fourier's law as long as the evolution rate is not very high. In nanosystems, the phonon-boundary scattering becomes important. The boundary scattering terminates the free travel of phonons, and shorten their MFPs. In a result, the effective thermal conductivity decreases. In regions far from the boundary, the MFPs of phonons are unaffected by the boundary. The boundary effect is limited in a region with the thickness proportional to the phonon MFPs. If the boundary affected region is much smaller than the characteristic size of the heat conduction, the boundary effect can be neglected. So the heat conduction process is independent of the system size in bulk materials. The boundary effect is nonnegligible if the boundary affected region is comparable with the system size, which induces the size effect of heat conduction in nanosystems.

To model such size effect, the boundary effect should be assessed properly. When the characteristic size of heat conducting system, l, is very small, the effective phonon MFPs, λ_R, is close to l. Thus the effective thermal conductivity is

$$\frac{\kappa_{\text{eff}}}{\kappa_0} = \frac{l}{\lambda_{R,0}}, \tag{6.2}$$

where the subscript 0 denotes the bulk value. Equation 6.2 is applicable for $l \ll \lambda_{R,\,0}$. In real experiments, l is in the same magnitude of $\lambda_{R,\,0}$. For example, it is recognized that the single-crystal Si has the MFP of 40–260 nm. The nanofilms and nanowires prepared in the experiments have the thickness larger than 20 nm. Apparent decrease of thermal conductivity can be observed when the system thickness reduces to 100 nm. To model this situation, the Matthiessen law is adopted to characterize the effect of boundary scattering on the phonon MFPs

$$\frac{1}{\lambda_{\text{eff}}} = \frac{1}{\lambda_{R,0}} + \beta\frac{1}{l}, \tag{6.3}$$

where β is the correction factor. The effective thermal conductivity of system is

$$\frac{\kappa_{\text{eff}}}{\kappa_0} = \frac{\lambda_{\text{eff}}}{\lambda_{R,0}} = \frac{1}{1+\beta\frac{\lambda_{R,0}}{l}} = \frac{1}{1+\beta Kn_R}. \tag{6.4}$$

Equation 6.4 is the gray model proposed by Majumdar [1]. $Kn_R = \lambda_{R,\,0}/l$ is the Knudsen number based on $\lambda_{R,\,0}$. The coefficient β depends on the geometry of heat conducting nanosystems. For the in-plane heat conductivity of nanofilms, $\beta = 3/8$. For the cross-plane heat conductivity of nanofilms, $\beta = 4/3$. For the longitudinal heat conductivity of nanowires, β can be selected as 4/3 [2]. Equation 6.4 indicates that the Knudsen number is large in nanosystems, which reduces the κ_{eff} from the bulk value. When the system thickness increases, Kn_R reduces, then the effective thermal conductivity approaches the bulk value.

The gray model gives qualitatively the dependence of the effective thermal conductivity on the system size. However, it has evident deviation from the experimental value quantitatively. It is claimed that the average MFP used in the gray model is based on the group velocity and specific heat of bulk material. In real cases, phonons of different frequencies contribute differently to the heat conduction. The low-frequency phonons have bigger group velocity and lower specific heat. Thus, they contribute mainly to the heat conduction. The high-frequency phonons contributes mainly to the specific heat, while have less contribution to the heat conduction. In this sense, the dispersion relation of phonons should be considered. Chen et al., [3–5] proposed that the phonon MFPs of single-crystal Si at room temperature should be 210–260 nm. If the dispersion relation is fully considered, more sophisticated prediction model can be obtained by integrating in the frequency domain. For instance, McGaughey et al., [2] proposed that the effective thermal conductivity can be expressed as:

$$\kappa_{\text{eff}} = \frac{3}{8\pi^2}\int_0^{\omega_D}\int_0^{2\pi}\int_0^{\pi} k_B \frac{v_{g,i}^2}{v_s^3}\tau\omega^2 \sin\theta \, d\theta d\varphi d\omega, \tag{6.5}$$

where ω_D is the Debye frequency. For simplicity it can be assumed that the phonon dispersion relation is linear, and only one phonon branch is considered. $v_{g,\,i}$ is the effective group velocity obtained in different geometries. McGaughey et al., [2] thus developed a prediction model for the in-plane and cross-plane effective thermal conductivity of nanofilms and the longitudinal conductivity of nanowires. This model matches well with experiments for nanofilms, while still overestimates the experiments for nanowires. The above-mentioned models evolve beyond the gray approximation by Majumdar. Nevertheless, the Matthiessen law is adopted by all of the models to account the MFP decrease due to the boundary effect. Therefore, they have no essential difference with the gray model. On the other hand, the detailed consideration of phonon dispersion relation requires more fitting parameters, which increases the arbitrary of the results.

Based on the EIT theory, Alvarez and Jou [6] elucidated that the Fourier heat conduction law can be written into the Fourier series. In nanosystems, the phonon wave package is assumed to form a series of standing waves with the wavelength the fraction of characteristic size of systems. If all the phonons have the same relaxation time regardless of their frequencies, the effective thermal conductivity of nanofilms is expressed as:

$$\frac{\kappa_{\mathrm{eff}}}{\kappa_0} = \frac{1}{2\pi^2 \left(Kn_R\right)^2} \left[\sqrt{1 + 4\pi^2 \left(Kn_R\right)^2} - 1 \right]. \tag{6.6}$$

For other geometries, Alvarez et al. [7] proposed that the effective characteristic size can be expressed by the average of the 3 dimensional sizes

$$\frac{1}{l_{\mathrm{eff}}^2} = \frac{1}{l_x^2} + \frac{1}{l_y^2} + \frac{1}{l_z^2}. \tag{6.7}$$

For nanowires, Kn_R in Eq. 6.6 is

$$Kn_R^{\mathrm{nw}} = 2\sqrt{2}\frac{\lambda_{R,0}}{D}, \tag{6.8}$$

where D is the diameter of the nanowire. For the silicon nanofilm at room temperature, if κ_0 is chosen as 120 W/(m K), $\lambda_{R,\,0}$ is 40 nm, this model agrees well with the experimental results.

The above models have these common assumptions: (1) The boundary effect of nanosystems is accounted by the reduced MFPs of phonons due to boundary scattering. (2) Only the effect of R process on heat conduction is considered. The role of N process is not considered.

Guyer and Krumhansl [8, 9] derived the linear solution of the phonon Boltzmann equation and got the more general heat conduction model beyond the Fourier's law (GK model, cf. Eq. 1.7). For the steady heat conduction in the constant cross-section systems, the GK model is simplified as:

$$\frac{1}{3}\left(v_{\mathrm{s}}\right)^2 \nabla e = -\frac{\boldsymbol{q}}{\tau_{\mathrm{R}}} + \frac{\tau_{\mathrm{N}}\left(v_{\mathrm{s}}\right)^2}{5} \nabla^2 \boldsymbol{q}. \tag{6.9}$$

In phonon systems the Fourier's law can be written as:

$$\frac{1}{3}\tau_{\mathrm{R}}\left(v_{\mathrm{s}}\right)^2 \nabla e = \kappa \nabla T \tag{6.10}$$

With this Eq. 6.9 can be transformed as:

$$\kappa \nabla T = -\boldsymbol{q} + \lambda_{\mathrm{B}}^2 \nabla^2 \boldsymbol{q}, \tag{6.11}$$

where

$$\lambda_{\mathrm{B}} = \sqrt{\frac{\tau_{\mathrm{R}} \tau_{\mathrm{N}} v_{\mathrm{s}}^2}{5}} = \sqrt{\frac{\lambda_{R,0} \lambda_{N,0}}{5}}. \tag{6.12}$$

Compared with the Fourier's law, Eq. 6.11 has an additional term which is the second-order spatial derivative of heat flux. Remind that the heat flux is similar to a fluid flow flux, $\nabla^2 q$ is in analogy with the viscous dissipation term in Navier–Stokes equation [9].

In the steady heat conduction through a constant cross-section medium, if one assumes that the heat flux vanishes at the boundary, then $\nabla^2 q$ in Eq. 6.11 induces the nonuniform heat flux distribution in the cross section, which is in analogy to the velocity profile in the Poiseuille flow. This behavior is called the phonon hydrodynamics. Guyer and Krumhansl [9] analyzed the effect of $\nabla^2 q$ on the heat conduction. They indicated that the Poiseuille flow would only happen in condition of $\lambda_{N,\,0} \ll l$ and $\lambda_{N,\,0}\lambda_{R,\,0} \gg l$, which induces the reduction of effective thermal conductivity. This condition is supposed to be satisfied by a solid helium sample with a characteristic radius 1 cm and at an extremely low temperature (0.62–0.76 K). The nanoscale material was not available in 1960s, so the Poiseuille flow of phonons can only be observed at extremely low temperature in order to have large MFP. In recent years, the nano-sized material is abundantly synthesized and the effect of $\nabla^2 q$ can be observed even at room temperature.

Alvarez et al., [7, 10–13] investigated the size effect of the thermal conductivity in nanosystems based on Eq. 6.11. The λ_{B} in Eq. 6.11 is assumed as the bulk MFP of phonons, $\lambda_{R,\,0}$. The relative importance of $\nabla^2 q$ can be measured by Kn_R. $\nabla^2 q$ can be neglected when Kn_R is small, and Eq. 6.11 reduces to the traditional Fourier's law. When $Kn_R \gg 1$, the magnitude of the $\nabla^2 q$ term will be much larger than q. At this time, the heat flux profile approaches to the parabolic sharp of the Poiseuille flow. Consider the in-plane heat conductions in nanofilms and nanowires are shown in Fig. 6.1a, b. If one assumes the nonslip boundary condition

$$q\left(y = \pm\frac{l}{2}\right) = 0 \tag{6.13}$$

The spatial distribution of heat flux can be obtained. For nanofilms and nanowires, when $Kn_{\mathrm{R}} \gg 1$, the total heat fluxes passing through each cross section are [7, 10–13]

$$Q^{\mathrm{nf}} = -\frac{bl^3 \kappa_0}{12\lambda_{R,0}^2} \nabla T \tag{6.14a}$$

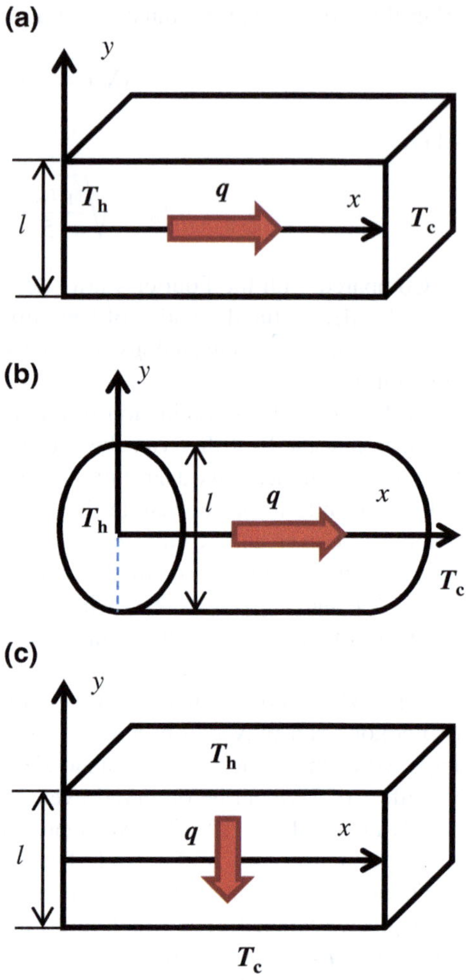

Fig. 6.1 Heat conduction in nanosystems. **a** In-plane nanofilm; **b** nanowire; **c** cross-plane nanofilm

$$Q^{\mathrm{nw}} = -\frac{\pi l^4 \kappa_0}{128\lambda_{R,0}^2}\nabla T, \tag{6.14b}$$

where b is the width of the nanofilm and assumed to be much larger than l to ensure the quasi-2D heat conduction, the superscripts, nf and nw, are the abbreviation of nanofilms and nanowires. Thus the corresponding effective thermal conductivities are

$$\kappa_{\mathrm{eff}}^{\mathrm{nf}} = \frac{l^2\kappa_0}{12\lambda_{R,0}^2} = \frac{\kappa_0}{12Kn_R^2}, \tag{6.15a}$$

$$\kappa_{\text{eff}}^{\text{nw}} = \frac{l^2 \kappa_0}{32 \lambda_{R,0}^2} = \frac{\kappa_0}{32 Kn_R^2}. \tag{6.15b}$$

Note that the Kn_R in Eq. 6.15a, 6.15b is defined based on the thickness of nanofilm and the diameter of nanowire. According to Eq. 6.15a, 6.15b, the effective thermal conductivity of nanosystems should be inversely proportional to the square of Kn_R due to the nonuniform distribution of heat flux profile. However, the experiments indicated that the effective thermal conductivity is approximately linear to the characteristic size rather than the square of size. It is thereby further elucidated that the boundary velocity slip would happen in case of large Kn_R. In analogy to the velocity slip at boundary for rarefied gases, the boundary heat flux can be written as [7, 10–13] :

$$q\left(y = \pm \frac{l}{2}\right) = -C \lambda_{R,0} \frac{\partial q}{\partial y}, \tag{6.16}$$

where C is the dimensionless slip coefficient. Then the effective thermal conductivities turn to

$$\kappa_{\text{eff}}^{\text{nf}} = \frac{\kappa_0}{12 Kn_R^2} (1 + 6 C Kn_R), \tag{6.17a}$$

$$\kappa_{\text{eff}}^{\text{nw}} = \frac{\kappa_0}{32 Kn_R^2} (1 + 8 C Kn_R). \tag{6.17b}$$

In case of $Kn_R \gg 1$, Eq. 6.17a, 6.17b predicts the linear size dependence of the effective thermal conductivity, which agrees with the trend observed in experiments. Ma [14] also studied the effective thermal conductivity of nanosystems based on Eq. 6.11. He proposed a slip boundary condition as:

$$q\left(y = \pm \frac{l}{2}\right) = q_0 \frac{\lambda_{\text{eff}}}{\lambda_{R,0}}, \tag{6.18}$$

where q_0 is the heat flux based on the Fourier's law

$$q_0 = -\kappa_0 \nabla T \tag{6.19}$$

λ_{eff} is the effective MFP based on the Matthiessen's law

$$\frac{\lambda_{\text{eff}}}{\lambda_{R,0}} = \frac{1}{1 + Kn_R (3 - j)^{-1}}, \tag{6.20}$$

where $j = 1$ for nanofilms, $j = 2$ for nanowires. Assume that $\lambda_{R,\,0} = 210$ nm, Ma's model obtains better prediction than the McGaughey's model [2]. However, Ma's model still has considerable discrepancy with the experiments for nanowires.

Compared with Eq. 6.17a, 6.17b, Ma's model excludes the slip parameter C. Thus this model cannot take into account the potential variation of boundary conditions, such as the roughness and reflectance.

The feature of the phonon hydrodynamics model is that the boundary effect is described by a viscous term as in fluid dynamics rather than the Matthiessen's law. This model can better characterize the geometry dependence of the effective thermal conductivity, namely, the difference between nanofilms and nanowires. Nevertheless, there are still some ambiguities in the present phonon hydrodynamics models. For example, the derivation of Eq. 6.11 contains the effect of N processes. λ_B is the product of λ_R and λ_N. The present phonon hydrodynamic models simply appoint λ_B to λ_R as the characteristic MFP. On the other hand, Kn_R is large in nanosystems. The rarefication effect on the effective MFP should be properly considered. These issues will be addressed in the following sections.

6.2 Phonon Gas Dynamics Based on Thermomass Theory

6.2.1 Viscosity of Phonon Gas

The fluid flow in porous medium is described by the Darcy's law

$$-\nabla p = \frac{\mu}{K} u_{\mathrm{f}}, \tag{6.21}$$

where K is the permeability with a unit of m^2. Based on the Darcy's law, the velocity profile in each cross section is uniform. The Darcy's law is applicable in case of small flow velocity, large scale flow, and thereby the boundary effect is negligible. When the investigated flow region is close to the boundary of porous medium, the boundary effect should be considered. Beavers and Joseph [15] studied the boundary condition at the interface of porous material and free flow. They indicated that the flow has a slip boundary condition at the interface. The boundary slip velocity attenuates from the boundary with a characteristic length of $K^{1/2}$ to the uniform velocity in the porous medium. Similarly, if the porous flow satisfies the nonslip boundary condition at the fixed walls, the Darcy's law should also be modified. Brinkman [16], [17] proposed that a second-order spatial derivative term can be added into the Darcy's law, making it in the same order as the governing equations for free flow, i.e.,

$$-\nabla p = \frac{\mu}{K} u_{\mathrm{f}} - \mu \nabla^2 u_{\mathrm{f}}. \tag{6.22}$$

Equation 6.22 is called the Brinkman extension to the Darcy's law. The flow satisfying Eq. 6.22 is called the Darcy–Brinkman flow.

In Sect. 2.2, it is pointed out that the friction force on the phonon gas should contain the second-order spatial derivative term of the drift velocity when the boundary effect is important

$$\boldsymbol{f}_{\mathrm{h}} = -\chi\rho_{\mathrm{h}}\boldsymbol{u}_{\mathrm{h}} + \mu_{\mathrm{h}}\nabla^2\boldsymbol{u}_{\mathrm{h}} \tag{2.24}$$

Remind that $f_{\mathrm{h}} = \nabla p_{\mathrm{h}}$ in steady flow (conduction). Comparing Eq. 2.24 with Eq. 6.22, one gets the permeability of the heat conductive medium

$$K_{\mathrm{h}} = \frac{\mu_{\mathrm{h}}}{\chi\rho_{\mathrm{h}}}. \tag{6.23}$$

Based on Eq. 2.24, the governing equation in the steady, constant cross-section heat conduction can be written as:

$$-\kappa\nabla T = q - K_{\mathrm{h}}\nabla^2 q = q - l_{\mathrm{B}}^2\nabla^2 q. \tag{6.24}$$

It is a simplified form of the general heat conduction law, Eq. 2.25, by reserving only the driving force and friction force. Note that l_{B} in Eq. 6.24 is extracted from the macroscopic derivation. Its square equals the permeability of the phonon gas flow in the material. The permeability is a macroscopic quantity proportional to the viscosity of phonon gas.

If Eq. 6.24 is regarded to be equivalent to Eq. 6.11, then $\lambda_{\mathrm{B}} = l_{\mathrm{B}}$. These two quantities can be understood as the macroscopic and microscopic descriptions of the same physical behavior. In fluid dynamics, the viscous term in Navier–Stokes equation can be derived from the first-order Chapman–Enskog expansion of the state distribution function of fluid molecular. Thus the distribution function, f, can be expressed as:

$$f = f_0 + Knf_1, \tag{6.25}$$

where Knf_1 is the first-order expansion term, Kn is the Knudsen number, representing the ratio of the MFP of fluid molecular over the characteristic size of system. Similarly, it is shown in Sect. 2.3 that the $\nabla^2 q$ term in Eq. 6.11 can be derived from the first-order Chapman–Enskog expansion of the state distribution function of phonons. In fluid dynamics, if the system size is large enough and thereby $Kn \ll 1$, Knf_1 can be neglected compared with f_0. In this case, the derivation based on the Boltzmann equation gives the Euler equation, which is the dynamic equation without the viscous dissipation. The large system condition can be also understood as the interested flow region is far away from the boundary. For instance, the Euler equation is suitable to describe gas flow far from the aircrafts in aerodynamics.

The λ_{B} in Eq. 6.11 is the function of both λ_R and λ_N. Hence it represents the average effect of the N and R scatterings of phonons. Equation 6.9 can be transformed as

$$\frac{1}{3}v_{\mathrm{s}}\nabla e = -\frac{\boldsymbol{q}}{\lambda_{\mathrm{R}}} + \frac{\lambda_{\mathrm{N}}}{5}\nabla^2\boldsymbol{q}. \tag{6.26}$$

Based on Eq. 1.12, λ_R is proportional to the thermal conductivity

$$\kappa = \frac{\rho C_V v_{\mathrm{s}}}{3}\lambda_R. \tag{1.12}$$

Inserting it into Eq. 6.26 gives

$$-\nabla T = \frac{\boldsymbol{q}}{\kappa} - \frac{3\lambda_{\mathrm{N}}}{5v_{\mathrm{s}}\rho C_V}\nabla^2\boldsymbol{q}. \tag{6.27}$$

Remind that

$$-\nabla p_{\mathrm{h}} = \frac{-2\gamma_{\mathrm{G}}\rho C_{\mathrm{V}}^2 T}{c^2}\nabla T. \tag{6.28}$$

Compare the derivation based on thermomass theory and phonon Boltzmann equation, τ_{TM} can be regard to be equivalent to τ_{R}, which means

$$2\gamma_{\mathrm{G}} C_{\mathrm{V}} T = \frac{1}{3}v_{\mathrm{s}}^2. \tag{6.29}$$

In this case, Eq. 6.28 turns to

$$-\nabla p_{\mathrm{h}} = -\frac{1}{3}\frac{\rho C_{\mathrm{V}} v_{\mathrm{s}}^2}{c^2}\nabla T. \tag{6.30}$$

Thus in the steady, constant cross-section conduction, the governing equations of thermomass motion are written as:

$$\begin{aligned} -\nabla p_{\mathrm{h}} &= \frac{\mu_{\mathrm{h}}}{K_{\mathrm{h}}}\boldsymbol{u}_{\mathrm{h}} - \mu_{\mathrm{h}}\nabla^2\boldsymbol{u}_{\mathrm{h}} \\ -\nabla T &= \frac{3c^2}{\rho C_V v_{\mathrm{s}}^2}\left(\frac{\mu_{\mathrm{h}}}{K_{\mathrm{h}}}\frac{q}{\rho C_V T} - \mu_{\mathrm{h}}\nabla^2\frac{q}{\rho C_V T}\right) \end{aligned}. \tag{6.31}$$

Compare it with Eq. 6.27, one gets

$$\kappa = \frac{K_{\mathrm{h}}}{\mu_{\mathrm{h}}}\frac{\rho^2 C_{\mathrm{V}}^2 T v_{\mathrm{s}}^2}{3c^2}, \tag{6.32a}$$

$$\mu_{\mathrm{h}} = \frac{\rho C_{\mathrm{V}} v_{\mathrm{s}} T}{5c^2}\lambda_N, \tag{6.32b}$$

where

$$K_{\mathrm{h}} = \lambda_{\mathrm{B}}^2 = \frac{\lambda_R \lambda_N}{5}. \tag{6.33}$$

Based on Eqs. 6.32a, 6.32b, and 6.33 one can also recover Eq. 1.12. Equation 6.32a, 6.32b shows that the phonon viscosity coefficient, μ_{h}, is proportional to λ_N, and does not depend on λ_R. Therefore, Eq. 6.27 can be reformed as:

$$-\nabla T = \frac{\boldsymbol{q}}{\kappa} - \frac{3c^2}{\left(\rho C_V v_{\mathrm{s}}\right)^2 T} \mu_{\mathrm{h}} \nabla^2 \boldsymbol{q} = \frac{\boldsymbol{q}}{\kappa} - \frac{1}{\rho^2 C_V^2 \xi T} \mu_{\mathrm{h}} \nabla^2 \boldsymbol{q}. \tag{6.34}$$

Compared with Eqs. 6.26, 6.34 uses the macroscopic variables such as κ and μ_{h}, instead of the microscopic variables relating to the phonon scattering like λ_N and λ_R. In terms of the phonon gas dynamics, the permeability K_{h} is more essential than the thermal conductivity, κ. Nonetheless, κ is adopted here for the convenience to compare the current model to the Fourier's law and experimental results.

Analysis based on Eqs. 6.11 and 6.24 indicates that the relative magnitude of λ_N, λ_R and l affects the feature of heat conduction. If

$$K_{\mathrm{h}} = \lambda_{\mathrm{B}}^2 = \frac{\lambda_R \lambda_N}{5} \ll l^2 \tag{6.35}$$

which means that the system size is far larger than the phonon MFPs, the second term on the right-hand side of Eq. 6.24 is much less than the first term, leading to the recovery of the Fourier's law. In contrast, when the system size is comparable to the phonon MFPs, both terms on the right-hand side of Eq. 6.24 contribute to the heat conduction. The solution of Eq. 6.24 gives the flux profile on each cross section. Then the effective thermal conductivity is obtained.

The *N* processes are regarded to conserve the phonon quasi-momentum over scattering. Hence the changes related to *N* processes are momentum conservative. Similarly, the momentum transport induced by the viscous effect in fluid dynamics is also momentum conservative. Generally the nonslip boundary condition will absorb the momentum at the boundary. Therefore in the typical Poiseuille flow the momentum loss happens at the boundary. In heat conduction, if λ_R is much larger than λ_N, namely, the frequency of *R* processes is rare, the momentum conservative *N* processes are prominent. The collective behavior of phonon gas is thus close to a flow of continuous fluid in a channel.

The *R* processes break the phonon quasi-momentum. It can be regarded as the collision of phonon on some obstacle and lose its momentum. Thus the *R* process dominated heat conduction is in analogy to the porous flow. The momentum loss happens inside the flow region. When the system size is large, the boundary effect is negligible. The momentum loss induced by the porous medium is prominent. In this case, the flow is diffusive. Correspondingly, the traditional Fourier's law considers only the effect of R processes. So it is called the diffusive heat conduction.

In the fully developed laminar flow, the boundary effect (friction) spreads among the whole flow region. The thickness of boundary layer equals the size of flow region. In porous medium, the viscous effect is prominent near the boundary, leading to the nonuniform velocity distribution on the cross section. Away from the boundary, the boundary effect attenuates with a characteristic length, l_B, and the velocity profile tends to be uniform. With a long enough distance, the boundary effect can be neglected. It indicates that the thickness of boundary layer is proportional to l_B in the porous flow. When the distance between the interested region and the boundary is comparable with or less than l_B, the boundary effect should be accounted. If the size of whole flow region is comparable with l_B, the velocity profile of flow should be characterized by the Darcy–Brinkman model.

A dimensionless number, Br, can be defined for the Darcy–Brinkman flow

$$\mathrm{Br} = l_\mathrm{B}/l \tag{6.36}$$

For the fully developed flow, the heat flux profile in nanofilms (Fig. 6.1a) derived from Eq. 6.24 is

$$q(y) = -\kappa_0 \nabla T\left[1 - \frac{\cosh(y/l_\mathrm{B})}{\cosh(l/2l_\mathrm{B})}\right]. \tag{6.37}$$

The corresponding effective thermal conductivity is

$$\kappa_\mathrm{eff}^\mathrm{nf} = \frac{\int_l q\mathrm{d}y}{-\nabla T l} = \kappa_0[1 - 2\mathrm{Br}\cdot\tanh(1/2\mathrm{Br})]. \tag{6.38}$$

For the circular cross section, i.e., nanowires (Fig. 6.1b), it is obtained from Eq. 6.24 that

$$q(y) = -\kappa_0 \nabla T\left[1 - \frac{J_0(iy/l_\mathrm{B})}{J_0(il/2l_\mathrm{B})}\right], \tag{6.39}$$

where J is the cylindrical Bessel function

$$J_n(x)=\left(\frac{x}{2}\right)^n\sum_{t=0}^{\infty}\frac{(-1)^t(x/2)^{2t}}{t!(t+n)!}. \tag{6.40}$$

The effective thermal conductivity is

$$\kappa_\mathrm{eff}^\mathrm{nw} = \kappa_0\left[1 - (4\mathrm{Br})\cdot\frac{J_1(i/2\mathrm{Br})}{iJ_0(i/2\mathrm{Br})}\right] = \kappa_0\left[1 - \frac{\sum_{t=0}^{\infty}\frac{(4\mathrm{Br})^{-2t}}{t!(t+1)!}}{\sum_{t=0}^{\infty}\frac{(4\mathrm{Br})^{-2t}}{t!t!}}\right]. \tag{6.41}$$

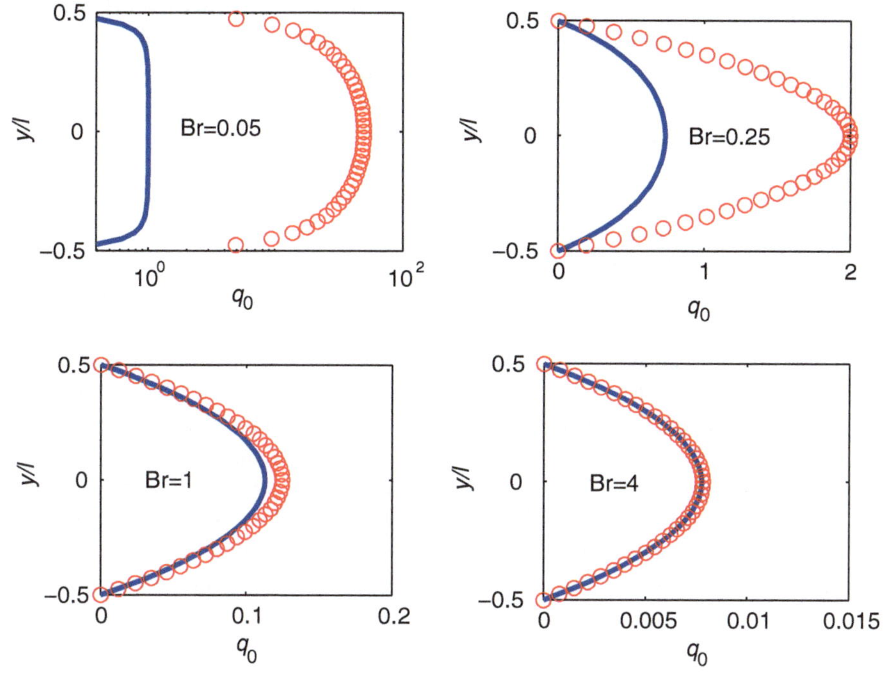

Fig. 6.2 Heat flux profiles at different Br in two-dimensional rectangle channels. $q_0 = -\kappa_0 \nabla T$ is the bulk heat flux. *Solid line* Darcy–Brinkman model; *Circle* Navier–Stokes model. Reprinted from size-dependent thermal conductivity of Si nanosystems based on phonon gas dynamics, Vol 56, Dong Y, Cao BY, Guo ZY, 256–262. Copyright (2014), with permission from Elsevier

For a 2D rectangle channel, the heat flux profiles at different Br are shown in Fig. 6.2. Comparison is made to the Navier–Stokes model, which only consider the phonon viscous term (second term on the right-hand side of Eq. 6.24) and neglect the porous friction term (first term on the right-hand side of Eq. 6.24). It can be seen from Fig. 6.2 that the average heat flux predicted by the Navier–Stokes model is much larger than q_0 for small Br. Based on the Darcy–Brinkman model, there is a boundary layer near the boundary, while the heat flux is uniform in the central region, equals q_0. For large Br, the profiles based on the Navier–Stokes model get close to those based on the Darcy–Brinkman model. The difference of effective thermal conductivity is 9.1 % for Br = 1 and 0.6 % for Br = 4.

6.2.2 *Rarefication Effect of Phonon Gas*

Macroscopically, l_B represents the characteristic attenuation length of the boundary effect in porous flow. If l_B is much larger than the MFP of molecular, the parameters in Eq. 6.22 do not depend on the system size. Thus the solution of Eq. 6.22 is also size independent. For the phonon gas flow in nanosystems, the phonon MFPs could be comparable with the system size. The phonon-boundary collision will reduce the effective MFP. Thus, the parameters in Eq. 6.24 still depend on the system size. It is well known that the rarefication effect rises for the gas flow in micro-channels. The molecular-boundary collision reduces the effective viscosity, which is proportional to the gas MFP. Thus to assess the conductivity of the micro-channel, one should consider not only the nonuniform velocity profile of gas flow induced by the viscosity, but also the effective viscosity variation due to the rarefication effect, which is size dependent. Similarly, the viscosity of phonon gas, $\mu_{\rm h}$, is proportional to λ_N. Therefore, the rarefication effect of phonon gas should be considered in condition that λ_N is comparable with the system size. Meanwhile, the effective value of λ_R will also reduce when it is comparable with the system size. Such reduction leads to the decrease of the first term on the right-hand side of Eq. 6.26, thereafter the effective κ in Eq. 6.24 also decreases. Remind that the κ discussed here is not the out coming effective thermal conductivity of the system. Instead it is merely a coefficient proportional to λ_R. Therefore, the λ_N and λ_R on the right-hand side of Eq. 6.26 reflect separately the two effects in phonon gas flow. In nanosystems, these two effects should be considered individually. In terms of the porous flow, the reduction of λ_N means the decrease of the effective viscosity, while the reduction of $\lambda_R \lambda_N$ is equivalent to the decrease of the permeability.

If the boundary is assumed to be pure diffusive, namely, the emitted phonons are only determined by the boundary temperature and have no memory of the absorbed phonons. In this case, the boundary scattering can be regarded to terminate the phonon MFPs. Thus the MFPs of phonons can be localized. In bulk materials, the phonon MFP is λ_0 (denoting $\lambda_{R,\,0}$ or $\lambda_{N,\,0}$), then the probability of traveling a distance of x between successive collision for a phonon is

$$P(x, x + \Delta x) = \exp\left(-\frac{x}{\lambda_0}\right)\Delta\frac{x}{\lambda_0}. \tag{6.42}$$

In one-dimensional condition, the integral of P is

$$\int\limits_0^x P\mathrm{d}x = \int_0^x \exp\left(-\frac{x}{\lambda_0}\right)\mathrm{d}\frac{x}{\lambda_0} = 1 - \exp\left(-\frac{x}{\lambda_0}\right). \tag{6.43}$$

It means the total possibility of scattering until the phonon travels to x. When $x \to \infty$ the integral is unity. If the boundary is located at r away from the investigated point, the effective MFP of phonons can be expressed as:

$$\frac{\lambda_{\rm eff}}{\lambda_0} = \int_0^r P\mathrm{d}x = 1 - \exp\left(-\frac{r}{\lambda_0}\right). \tag{6.44}$$

When $r \gg \lambda_0, \lambda_{\rm eff} \to \lambda_0$. It means that if the boundary is far away, the local effective MFP recovers the bulk value. Equation 6.44 is equivalent to the case that the investigated point locates at the center of a sphere with its radium r. For practical nanowires and nanofilms, the effective MFPs need to be integrated over the sphere angle.

For the 2D rectangle cross section (nanofilm) shown in Fig. 6.1a, Stops et al., [18] obtained the local effective MFP as:

$$\frac{\lambda_{\text{eff}}(y)}{\lambda_0} = 1 + \frac{1}{2}\left[(\alpha - 1)e^{-\alpha} + (\beta - 1)e^{-\beta} - \alpha^2 E_i(\alpha) - \beta^2 E_i(\beta)\right], \tag{6.45}$$

where

$$\begin{aligned} \alpha &= \frac{l/2 - y}{\lambda_0} \\ \beta &= \frac{l/2 + y}{\lambda_0} \\ E_i(x) &= \int_1^{\infty} t^{-1} e^{-tx} \mathrm{d}t \end{aligned} . \tag{6.46}$$

For the circular cross section (nanowire) shown in Fig. 6.1b, the local effective MFP is

$$\frac{\lambda_{\text{eff}}(y)}{\lambda_0} = \frac{1}{\pi}\int_0^{\pi}\int_0^{\frac{\pi}{2}}\left[1 - \exp\left(-\frac{y\cos\theta + \sqrt{l^2/4 - y^2\sin^2\theta}}{\lambda_0 \sin\varphi}\right)\right]\sin\varphi \mathrm{d}\varphi \mathrm{d}\theta. \tag{6.47}$$

The local effective MFP profiles at different Kn are shown in Fig. 6.3. It can be seen that the effective MFP is lower at the near boundary region, which is caused by the higher phonon-boundary scattering rate. In case of small Kn, the reduction of effective MFP only happens in near boundary region, with a characteristic thickness λ_0. The confinements on phonon MFPs by nanowires and nanofilms are close to each other in this condition. For large Kn, the reduction of MFP happens all over the flow region. Meanwhile, the confinement by nanowires is apparently stronger than that by nanofilms. When $Kn \gg 1$, the effective MFPs in the whole cross section of nanofilms approach to l.

With the effective MFPs, the heat flux profile can be calculated from Eq. 6.26 and the effective thermal conductivity is obtained. For convenience, Eq. 6.26 is rewritten as:

$$-\frac{\lambda_{R,\text{eff}}(r)}{\lambda_{R,0}}\kappa_0 \nabla T = \boldsymbol{q}(r) - \frac{\lambda_{R,\text{eff}}(r)\lambda_{N,\text{eff}}(r)}{5}\nabla^2 \boldsymbol{q}(r). \tag{6.48}$$

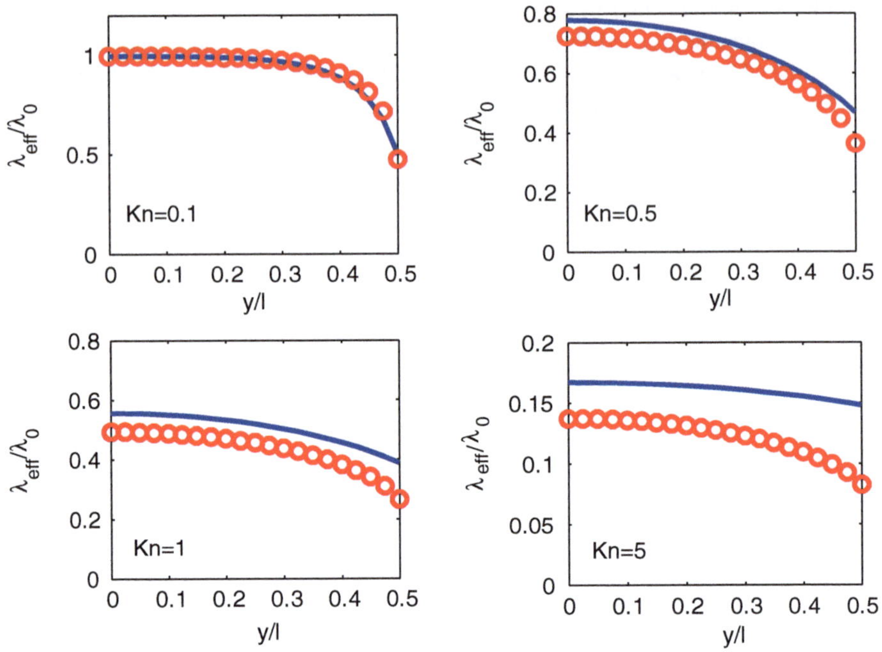

Fig. 6.3 Local effective MFP at different Kn based on Eqs. 6.45–6.47. *Solid line* Nanofilm; *Circle* Nanowire

The boundary condition needs to be specified to solve Eq. 6.48. In literature, the proposed boundary conditions include the Maxwell slip boundary (cf. Eq. 6.16) [7, 10–13] and modified MFP-based boundary (cf. Eq. 6.20) [14], etc. In these models, the constant MFP is used without considering the effective MFP reduction due to boundary scattering. In terms of the rarefied gas dynamics, these models only consider the boundary slip and neglect the change of viscosity. Equation 6.48 models the reduction of effective MFPs, which is equivalent to describing the rarefication effect by the reduction of viscosity. Thus, these two opinions can be regarded as modeling the same behavior from different perspectives. On the other hand, the numerical studies can fit the experiments with pure diffusive boundary. Thus the nonslip boundary condition can be used to solve Eq. 6.48 as long as the reduction of effective MFP is accounted.

6.3 In-Plane Thermal Conductivity of Si Nanosystems

Equation 6.48 is capable to describe the heating conduction with the temperature drop perpendicular to the characteristic size, namely, in-plane heat conduction (This term is also used for the longitudinal conduction in nanowires, which also satisfies the above criterion). There are plenty of experimental results for the single-crystal Si nanofilms [19–22] and nanowires [23]. Thus the effectiveness of Eq. 6.48 is evaluated with the experiments of Si nanosystems [19–23] as well as some numerical simulation results [24, 25]. The commonly adopted physical properties of single-crystal Si at room temperature are: κ_0 = 148 W/(m K), ρ = 2330 kg/m^3, C_V = 707 J/(kg K), v_s = 6400 m/s. Based on these properties, $\lambda_{R,\ 0}$ = 42 nm according to Eq. 1.12. The value of $\lambda_{N,\ 0}$ is not exactly reported in literature. The calculation based on the first principle indicated that $\lambda_{N,\ 0}$ is larger than $\lambda_{R,\ 0}$ at room temperature [26]. Here the numerical calculation shows that Eq. 6.48 gives good prediction for both nanofilms and nanowires with $\lambda_{N,\ 0}$ = 360 nm, as shown in Fig. 6.4. For comparison, the predictions based on the gray model [1] and McGaughey's model [2] are also presented in Fig. 6.4. The MFP used in the gray model is $\lambda_{R,\ 0}$ = 42 nm. It shows that the gray model and McGaughey's model both overestimate the effective thermal conductivity of nanowires. The prediction based on Eq. 6.48 is good for both nanofilms and nanowires. If $\nabla^2 q$ in Eq. 6.48 is neglected, the prediction will be close to that based on the gray model. Thus the main distinction of the present model with other models is considering the viscous effect of phonons. Without the viscous effect, one can hardly obtain good prediction for both nanofilms and nanowires (Fig. 6.5).

Besides the prediction at room temperature, the predictions at other temperatures for nanofilms and nanowires are also made. The bulk MFP of R processes, $\lambda_{R,\ 0}$, is achieved from [7, 10–13]. The optimal value of $\lambda_{N,\ 0}$ is obtained by fitting experiments. The values of κ_0, $\lambda_{R,\ 0}$, $\lambda_{N,\ 0}$ are listed in Table 6.1.

The value of $\lambda_{N,\ 0}$ is mainly based on the experiments of nanowires with a diameter 115 nm. For other sized nanowires, the boundary slip coefficient F is introduced as [27]. The values of F are 1.35, 1.18, and 1.0 for diameters of 56, 37, and 22 nm. F is 1.2 and 1.3 for nanofilms with thicknesses 100 and 20 nm. The thickness used in numerical modeling is Fl.

The accuracy is not high due to the fitting of $\lambda_{N,\ 0}$. Nevertheless, only one modification coefficient F is used for the nanosystems with a size (thus for a specified sample only one F is needed). The modification coefficient F is also adopted by Mingo [27] to predict the experiments. F is 1.15, 1.3, and 1.05 for nanowires with diameters 115, 56, and 37 nm, which are close to the present values. Mingo [27] proposed that if F is close to unity, it means that the boundary of nanowires is diffusive, which agrees with the nonslip boundary condition used in the present model.

The trends of temperature dependence of phonon MFPs are shown in Table 6.1. The $\lambda_{R,\ 0}$ continuously decrease when the temperature rises. This behavior is induced by the increased rate of phonon Umklapp scattering at higher temperatures due to the increased population of high-frequency (energy) phonons. The MFPs of N processes first decrease and then increase with temperature. The explanation could be: The total number of phonons is rare at low temperature. The frequency of N processes is low. As the temperature rises, the number of phonons increases. These phonons are mainly low energy. Then at this temperature range the N processes is prominent, with a short MFP. When the temperature further

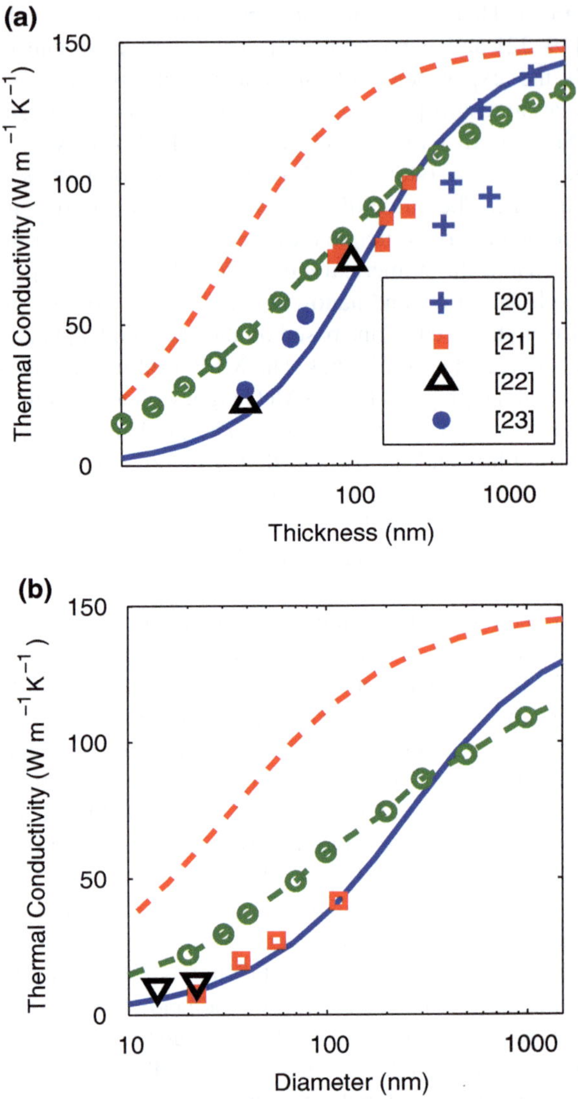

Fig. 6.4 Effective in-plane thermal conductivities of single-crystal Si nanosystems at room temperature. *Solid line* present model (Eq. 6.48); *Dash line* Gray model [1]; *Dash* and *circle* McGaughey model [2]. **a** Nanofilm. *Symbols* are experimental results from [19–22]. **b** Nanowire. *Square* is experimental results from [23]. *Triangle* is numerical results from [24, 25]

increases, the energy of phonons becomes large. Then the *R* processes grow, and the *N* processes reduce due to competition, leading to a larger $\lambda_{N,\,0}$.

It is found that the ratio of $\lambda_{R,\,0}$ over $\lambda_{N,\,0}$ approximately satisfies

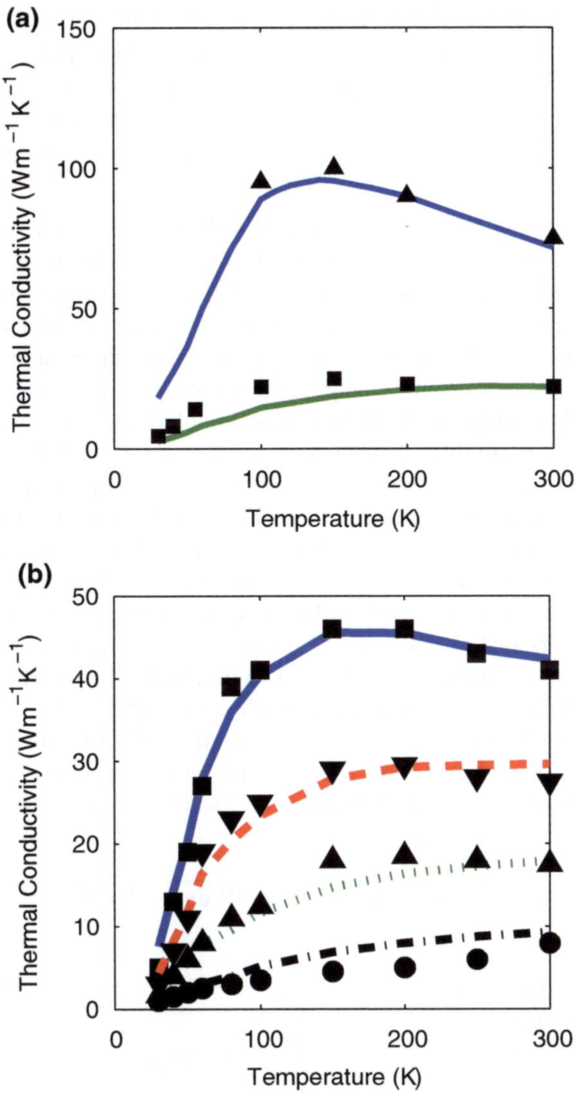

Fig. 6.5 Temperature-dependent effective in-plane thermal conductivities of single-crystal Si nanosystems. **a** Lines from *top* to *bottom*: predictions for nanofilms with thicknesses 100 and 20 nm. *Dots* Experimental results from [21]. **b** Lines from *top* to *bottom* predictions for nanowires with diameters 115, 56, 37 and 22 nm. *Dots* Experimental results from [23]

$$\frac{\lambda_{R,0}}{\lambda_{N,0}} = 0.0018\left(\frac{T_{\rm D}}{T}\right)^5 \tag{6.49}$$

Table 6.1 Temperature-dependent properties of single-crystal Si [7, 10–13]

T(K)	30	40	50	60	80	100	150	200	250	300
κ_0(W m^{-1} K^{-1})	4810	3530	2680	2110	1340	884	409	264	191	148
$\lambda_{R,\ 0}$(nm)	16,354	11,571	6681	3837	1432	557	180	93	58	42
$\lambda_{N,\ 0}$(nm)	50	10	7.5	7	16.5	45	75	135	260	360

which is shown in Fig. 6.6. Here $T_D = 645$ K is the Debye temperature of Si. Li [23] pointed out that the effective thermal conductivities of nanowires with diameters 115 and 56 nm at the temperature range 30–100 K are approximately proportional to T^3. This behavior is explained as: due to the size limit, the MFPs of phonons are nearly constant as the diameter of wires at this temperature range. Therefore, the effective thermal conductivity changes due to the specific heat, which has a T^3 dependence at low temperature. However, for smaller nanowires, with diameters 37 and 22 nm, the temperature dependences seem to be T^2 and T, which cannot be explained with the above model. The values of $\lambda_{R,\ 0}$ in Table 6.1 have a T^{-3} dependence. Thus the T^3 dependence for 115 and 56 nm nanowires can be also explained by the variation of $\lambda_{R,\ 0}$ with temperature. For thinner nanowires, their characteristic sizes are close to $\lambda_{N,\ 0}$ ($\approx$ 20 nm) at 30–100 K. Hence the viscosity effect is important and induces the different temperature dependence.

The parameters in Eq. 6.48 change in space. Thus, Eq. 6.48 needs to be integrated numerically, which is inconvenient to use. Once the parameters in Eq. 6.48 are average values which is constant in each cross section, the analytical solution can be obtained. In rarefied gas dynamics, Veijola and Turowski [28] investigated the relation between the boundary stress and velocity gradient for Couette flow based on Boltzmann equation. The effective viscosity at large Kn is

$$\frac{\mu_{\text{eff}}}{\mu_0} = \frac{1}{1 + 2Kn + 0.2Kn^{0.788}\exp(-Kn/10)}. \tag{6.50}$$

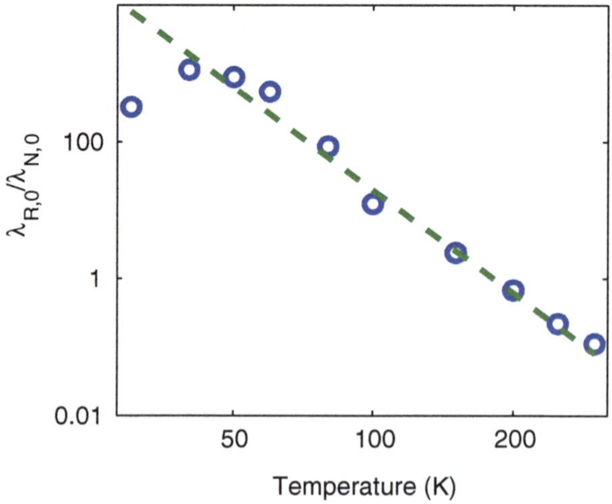

Fig. 6.6 Ratio of MFPs of R over N processes. *Circle* Table 6.1; *Dash line* Eq. 6.49

The effective viscosity is expressed by the ratio of the average shear stress over the velocity gradient at the boundary

$$\mu_{\mathrm{eff}} = -\frac{\langle \Pi_{xy} \rangle}{\langle \partial u / \partial y \rangle}. \tag{6.51}$$

The effective viscosity can be also obtained by the modification of MFP, i.e., Eq. 6.45. Guo et al., [29] pointed out that the local viscosity based on Eq. 6.45 can be inserted into the Navier–Stokes equation and predict well the velocity profile of rarefied gases. The total volume flux is also well predicted. Based on the relation between the flux and the viscosity in the steady planer Poiseuille flow

$$\mu_{\mathrm{eff}}(Kn) = -\frac{1}{12}\frac{l^3 W}{Q(Kn)}\frac{\partial p}{\partial x} \tag{6.52}$$

one can obtain the average viscosity. In Eq. 6.52 l is the planer distance, W is the width and Q is the volume flux. The dependences of the effective viscosity and Kn calculated based on Eqs. 6.50 and 6.52 are shown in Fig. 6.7. It shows that at high Kn the effective viscosities are linear to Kn^{-1}. In other words, the effective viscosity is proportional to the characteristic size at large Kn.

For phonon gas, the effective viscosity of nanosystems can be extracted from experiments based on Eqs. 6.38 and 6.41

$$\mu_{\mathrm{h,eff}} = \frac{2\gamma_{\mathrm{G}}\rho^2 C_{\mathrm{V}}^3 T^2}{c^2}\frac{l_{\mathrm{B}}^2}{\kappa_0} \tag{6.53}$$

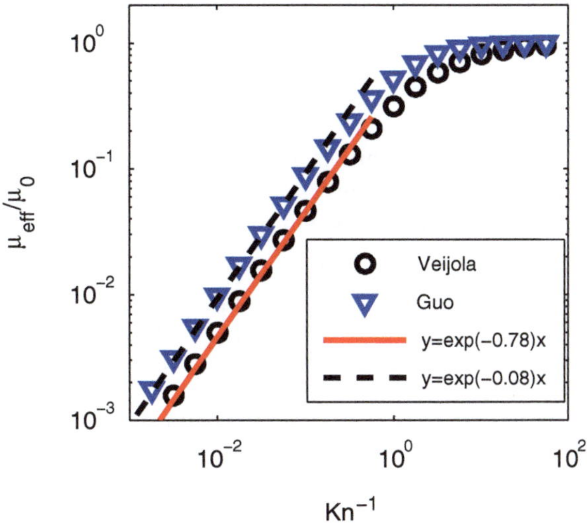

Fig. 6.7 Effective viscosity versus Kn based on Veijola model [28] (Eq. 6.50) and Guo model [29] (Eq. 6.52). Reprinted from size-dependent thermal conductivity of Si nanosystems based on phonon gas dynamics, Vol 56, Dong Y, Cao BY, Guo ZY, 256–262. Copyright (2014), with permission from Elsevier

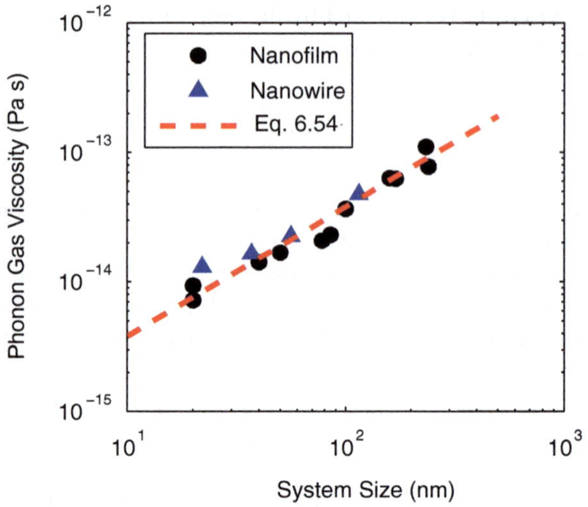

Fig. 6.8 Effective phonon gas viscosity extracted from experiments. *Circle dots* Nanofilms; *Triangles* Nanowires. *Dash line* Eq. 6.54. Reprinted from size-dependent thermal conductivity of Si nanosystems based on phonon gas dynamics, Vol 56, Dong Y, Cao BY, Guo ZY, 256–262. Copyright (2014), with permission from Elsevier

The size-dependent effective viscosity of Si at room temperature is shown in Fig. 6.8. Linear size dependence of viscosity is observed from Fig. 6.8, i.e.,

$$\mu_{\mathrm{h,eff}} = \varepsilon l, \tag{6.54}$$

where $\varepsilon = 3.83 \times 10^{-7}$ Pa s m^{-1}. As a result

$$l_{\mathrm{B}} = \sqrt{\frac{\kappa c^2 \varepsilon L}{2\gamma_{\mathrm{G}} \rho^2 C_{\mathrm{V}}^3 T^2}} = \sqrt{\lambda_{\mathrm{E}} L}, \tag{6.55}$$

where λ_E = 7.53 nm is the characteristic length. Substitution of Eq. 6.55 into Eqs. 6.38 and 6.41 gives the analytical expression of the effective thermal conductivity for nanofilms and nanowires

$$\kappa_{\mathrm{eff}}^{\mathrm{nf}} = \kappa_0 \left[1 - 2\sqrt{\frac{\lambda_{\mathrm{E}}}{L}} \cdot \tan h\left(\frac{1}{2}\sqrt{\frac{L}{\lambda_E}} \right) \right], \tag{6.56}$$

$$\kappa_{\mathrm{eff}}^{\mathrm{nw}} = \kappa_0 \left[1 - \frac{\sum_{t=0}^{\infty} \frac{(L/16\lambda_{\mathrm{E}})^t}{t!(t+1)!}}{\sum_{t=0}^{\infty} \frac{(L/16\lambda_{\mathrm{E}})^t}{t!t!}} \right]. \tag{6.57}$$

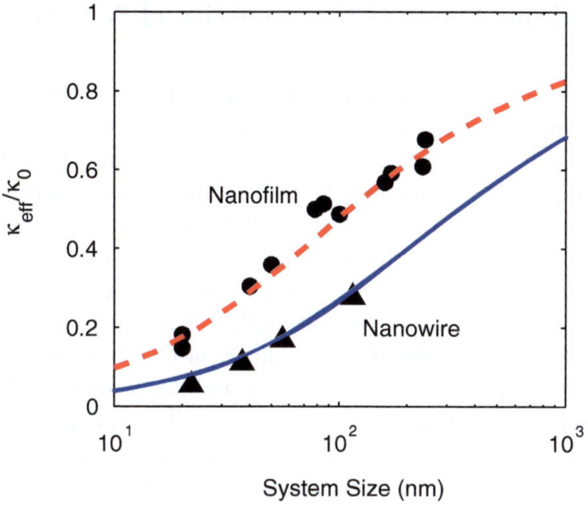

Fig. 6.9 Prediction of Eq. 6.56 (*Dash line*) and Eq. 6.57 (*Solid line*) versus experiments of nanofilms (*Circle dots*) and nanowires (*Triangles*). Reprinted from size-dependent thermal conductivity of Si nanosystems based on phonon gas dynamics, Vol 56, Dong Y, Cao BY, Guo ZY, 256–262. Copyright (2014), with permission from Elsevier

The prediction of Eqs. 6.56 and 6.57 agrees well with experiments, as shown in Fig. 6.9.

Equations 6.56 and 6.57 describe the phonon gas flow with the averaged parameters. The size-dependent effective viscosity reflects the rarefication effect. The form is analytical and convenient to use with reasonable accuracy.

The present models using phonon, phonon gas and MFPs belong to the perspective of particle motion, which is a discrete picture. In contrast, Eqs. 6.56 and 6.57 characterize the nanosystem heat conduction from the continuous perspective, namely, the viscosity of phonon gas. The relation between the macroscopic picture and microscopic picture can be discussed in analogy to the rarefied gas dynamics. According to Eq. 6.32b, the viscosity of phonon gas is proportional to λ_N. Figure 6.3 indicates that $\lambda_{N,\ \mathrm{eff}}$ consistently approaches to l when the system size is small. On the other hand, for Si at room temperature, $\lambda_{N,\ 0} \gg \lambda_{R,\ 0} \approx l$. Therefore, the viscosity reduction has significant impact on the thermal conductivity. At lower temperatures, $\lambda_{R,\ 0}$ is also large than the system size. The change of system permeability should also be accounted. In this case, the linear relation between the viscosity and system size could possibly be broken.

6.4 Cross-Plane Thermal Conductivity of Nanofilms

The cross-plane heat conduction in nanofilms refers to the condition that the temperature gradient is perpendicular to the face of films. There are some experiments measuring the cross-plane conductivity of superlattices, amorphous silicon nanofilms, and polycrystalline metals. Nevertheless, no experiments have been done for the cross-plane thermal conductivity (CPTC) of single-crystalline Si films. The small thickness of nanofilms causes the film resistance much less than the interfacial resistance, which impedes the current measurement method to obtain exactly the CPTC of Si nanofilms. The CPTC has been numerically calculated through the MD and MC methods. The numerical results show that the CPTCs of nanofilms are much less than the bulk value and decrease with the shrink of film thickness. For small film thickness, the effective CPTC is approximately linear to the film thickness, which indicates a constant thermal conductance in the cross-plane direction. Besides, in MC simulations, the significant jump of temperature is observed between the boundary and inner region. The CPTC of thin films has important application in the thermal management of IC chips and high-efficient thermal electrical materials based on nanotechnology.

Here the investigation of CPTC starts from the phonon Boltzmann equation

$$\left(\frac{\partial}{\partial t}+\boldsymbol{v}_k^n\cdot\nabla\right)f^n=\frac{f_0^n-f^n}{\tau_{\mathrm{R}}}+\frac{f_{\mathrm{D}}^n-f^n}{\tau_{\mathrm{N}}} \tag{2.32}$$

The phonon hydrodynamics model and the general heat conduction law based on thermomass theory have been derived based on this equation. However, the previous derivation has the following assumptions: (1) The temperature gradient is constant in the x direction, i.e., the system length is infinite. (2) The boundary condition of phonon distribution function is not specified, i.e., the distribution functions of inlet and outlet phonons do not vary significantly. These assumptions are equivalent to the full-developed condition in fluid dynamics. Therefore, the previous derivations are not applicable for the CPTC of nanofilms.

The phonon distribution function of phonons, f, could experience large variation during the cross-plane heat conduction. In near boundary regions, the phonons emitted from boundary are prominent. In this case, f is mainly determined by the boundary coefficients. The near boundary f could have big difference from the f far away from the boundary, thus the near boundary f can be defined as ballistic distribution, f_{B}, which is determined by the boundary emission. Leaving the boundary, the phonons will be continuously scattered in the medium, and f approaches to the bulk one. In the region far from the boundary, f should be fully relaxed to the bulk distribution. From Sect. 2.3, it shows that the distribution in bulk materials can be approximated by the displaced Planck distribution, f_{D}. (cf. Eq. 2.30). f_{D} represents the collective motion of phonon gas. On the other hand, at the boundary one has $f=f_{\mathrm{B}}$. Then Eq. 2.32 can be analyzed based on this boundary condition to obtain the model for CPTC of nanofilms.

In steady condition, Eq. 2.32 turns to

$$\boldsymbol{v}_k^n \cdot \nabla f^n = \frac{f_0^n - f^n}{\tau_R} + \frac{f_D^n - f^n}{\tau_N}. \tag{6.58}$$

Integration of Eq. 6.58 by multiplying the phonon momentum gives

$$-\int_k \hbar\omega^n \left(\boldsymbol{v}_k^n\right)^2 \cdot \nabla f^n = \int_k \left(\frac{f^n}{\tau_R} + \frac{f^n}{\tau_N} - \frac{f_D^n}{\tau_N}\right) \hbar\omega^n v_k^n. \tag{6.59}$$

The left-hand side of Eq. 6.59 corresponds to the temperature gradient. The right-hand side is proportional to the heat flux. The heat flux has the form

$$q = \sum_n \int_k \hbar\omega^n v_k^n f^n \tag{6.60}$$

Then Eq. 6.59 can be transformed as:

$$-\frac{1}{3}\rho C_V v_s^2 \nabla T = q\left(\frac{1}{\tau_R} + \frac{1}{\tau_N}\right) - \int_k \left(\frac{f_D^n}{\tau_N}\right) \hbar\omega^n v_k^n \tag{6.61}$$

In Sect. 2.3, it is assumed that $f \approx f_D$. Inserting it into Eq. 6.61 gives the Fourier's heat conduction law. Nevertheless, it is not the case in the cross-plane heat conduction of nanofilms. Instead, the boundary condition is $f = f_B$, i.e.,

$$q(x=0) = \sum_n \int_k \hbar\omega^n v_k^n f_B^n = q_B(x=0). \tag{6.62}$$

In the 1D geometry, the heat flux is the same at each cross section. Then in the whole system one has

$$q(x) = q_B(x) + q_D(x) = q_0. \tag{6.63}$$

Thus the phonon distribution function can be written as:

$$f^n(x) = f_B^n(x) + f_D^n(x) \tag{6.64}$$

Its integral with the phonon gas momentum is q_0 in each cross section. Equation 6.64 indicates that the phonon distribution function can be divided into two parts. One originates from the boundary emitted phonons, f_B. Another is the scattered distribution inside the medium, f_D. The phonon distribution function is close to f_B at the boundary. Then it gradually relaxes to f_D at a long enough distance from the boundary. The heat flux consists separate contributions from f_B and f_D. At the boundary the heat flux is fully contributed by f_B, while it is fully contributed by

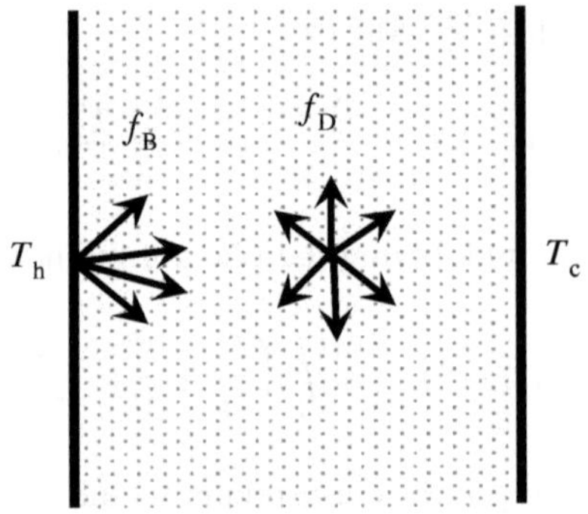

Fig. 6.10 Ballistic-diffusive transport in cross-plane heat conduction in nanofilms. Reprinted from Ballistic–diffusive phonon transport and size-induced anisotropy of thermal conductivity of silicon nanofilms, Vol 66, Dong Y, Cao BY, Guo ZY, 1–6. Copyright (2015), with permission from Elsevier

f_D far from the boundary. The sketch of this mechanism is shown in Fig. 6.10. Inserting Eq. 6.64 into Eq. 6.61 gives

$$\begin{aligned}
-\frac{1}{3}\rho C_V v_s^2 \nabla T &= q_0\left(\frac{1}{\tau_R}+\frac{1}{\tau_N}\right) - \int_k \left(\frac{f^n - f_B^n}{\tau_N}\right)\hbar\omega^n v_k^n \\
&= \frac{q_0}{\tau_R} + \int\int_k \left(\frac{f_B^n}{\tau_N}\right)\hbar\omega^n v_k^n \\
&= \frac{q_0}{\tau_R} + \frac{q_B}{\tau_N}
\end{aligned} \tag{6.65}$$

It shows that in regions far from the boundary, $q_B = 0$, and Eq. 6.65 reduces to the Fourier's law. At the boundary, $q_0 = q_B$, then the heat conduction satisfies

$$\begin{aligned}
-\frac{1}{3}\rho C_V v_s^2 \nabla T &= q_B\left(\frac{1}{\tau_R}+\frac{1}{\tau_N}\right) \\
-\frac{\kappa_0}{\lambda_{R,0}}\nabla T &= q_B\left(\frac{1}{v_s\tau_R}+\frac{1}{v_s\tau_N}\right)
\end{aligned}. \tag{6.66}$$

Equation 6.66 indicates that the ratio of q_B over the temperature gradient is the effective thermal conductivity, i.e.,

$$\begin{aligned}
-\kappa_{B,\mathrm{eff}}\nabla T &= q_B \\
\kappa_{B,\mathrm{eff}} &= \frac{1}{\lambda_{R,0}}\left(\frac{1}{\lambda_R}+\frac{1}{\lambda_N}\right)^{-1}\kappa_0
\end{aligned}. \tag{6.67}$$

In near boundary regions, the total heat flux q_0 is consisted of q_B and q_D. In this case the local temperature gradient drives q_B and q_D simultaneously. The effective thermal conductivity should be evaluated separately upon the contributions of q_B and q_D. Then Eq. 6.65 turns to

$$-\frac{\kappa_0}{\lambda_{R,0}}\nabla T = q_B\left(\frac{1}{\lambda_R}+\frac{1}{\lambda_N}\right)+\frac{q_D}{\lambda_R}. \tag{6.68}$$

Once the spatial distribution of q_B and q_D is obtained, the effective thermal conductivity can be derived based on the relation between the total heat flux and temperature drop.

According to Eq. 6.58, the N processes tend to relax f to f_D, with a relaxation time τ_N. Therefore, when f does not equal f_D, it will relax to f_0 due to the R scattering and meanwhile to f_D due to the N scattering. At the boundary, f has the largest deviation to f_D. Far away from the boundary f can be regarded to be close to f_D. Hence τ_N is the characteristic relaxation time from f_B to f_D, while λ_N is the characteristic relaxation length from q_B to q_D. In nanofilms, the boundary condition is that $q_D = 0$ at $x = \pm 0.5\ l$. q_B attenuates exponentially from the boundary with a characteristic length, λ_N. Based on these conditions, the distributions of q_B and q_D in the film are

$$q_B(x) = q_0\frac{\exp\left(-\frac{x}{\lambda_N}\right)+\exp\left(\frac{x}{\lambda_N}\right)}{\exp\left(-\frac{l}{2\lambda_N}\right)+\exp\left(\frac{l}{2\lambda_N}\right)} = q_0\frac{\cos h\left(\frac{x}{\lambda_N}\right)}{\cos h\left(\frac{l}{2\lambda_N}\right)}, \tag{6.69}$$

$$q_D(x) = q_0 - q_B(x) = q_0\left[1-\frac{\cosh\left(\frac{x}{\lambda_N}\right)}{\cos h\left(\frac{l}{2\lambda_N}\right)}\right]. \tag{6.70}$$

The heat flux profile under unit temperature drop across the film can be solved from Eqs. 6.68 to 6.70, which then gives the effective thermal conductivity. To calculate the CPTC of the single-crystal Si nanofilms, the properties are chosen the same as the in-plane geometry in the previous section, κ_0 = 148 W/(m K), $\lambda_{R,0}$ = 42 nm, $\lambda_{N,0}$ = 360 nm. Note that the local MFPs should be the effective MFPs, λ_{eff}, rather than the bulk value, λ_0. The relation between λ_{eff} and λ_0 has been derived in Eq. 6.45. The calculation results are shown in Fig. 6.11. The results based the gray model [1] and McGaughey model [2] are also presented for comparison. The evolution trends of different models are similar. The present model has a good agreement with the MD results [30, 31].

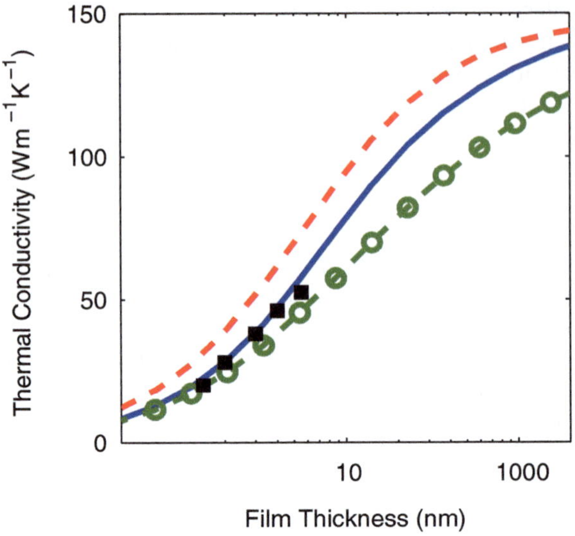

Fig. 6.11 Cross-plane thermal conductivity at room temperature. *Solid line* present model; *Dash line* gray model [1]; *Dash and circle* McGaughey model [2]; *Square Dots* MD results [30, 31]. Reprinted from ballistic–diffusive phonon transport and size-induced anisotropy of thermal conductivity of silicon nanofilms, Vol 66, Dong Y, Cao BY, Guo ZY, 1–6. Copyright (2015), with permission from Elsevier

A ballistic–diffusive model was also proposed by Chen [32]. Chen's model focuses on the heat wave propagation problem. Only *R* processes are considered in Chen's Boltzmann equation. Therefore, although the present model has a similar physical picture with Chen's work, the quantitative prediction is completely different.

With the modeling of the CPTC of nanofilms, one can further analyze the relation between the in-plane conductivity, $\kappa_{x,\text{ eff}}$, and the cross-plane conductivity, $\kappa_{y,\text{ eff}}$. The heterogeneity factor of the film's thermal conductivity is defined as:

$$A^{\mathrm{nf}} = \frac{\kappa_{y,\mathrm{eff}}}{\kappa_{x,\mathrm{eff}}} \tag{6.71}$$

The size dependence of A^{nf} of Si single-crystal film is shown in Fig. 6.12. The cross-plane conductivity is continuously larger than the in-plane conductivity. The heterogeneity is significant for thin films and gradually vanishes with the increase of film thickness.

Figure 6.12 indicates that the maximum heterogeneity factor is 1.8 when the film thickness approaches to zero. This limit can be derived theoretically. The effective MFPs are calculated through Eq. 6.45. When the film thickness approaches to zero, the effective MFPs, $\lambda_{R,\text{ eff}}$ and $\lambda_{N,\text{ eff}}$, all approach to the film thickness, *l*. For the

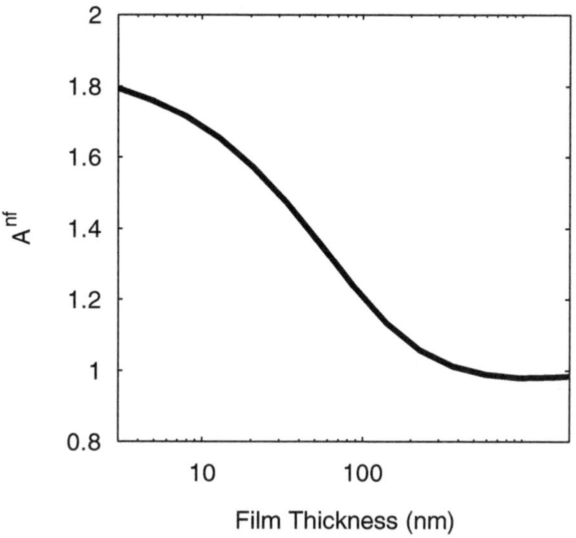

Fig. 6.12 Heterogeneity of thermal conductivity of Si nanofilm at room temperature. A^{nf} is given by Eq. 6.71. Reprinted from Ballistic–diffusive phonon transport and size-induced anisotropy of thermal conductivity of silicon nanofilms, Vol 66, Dong Y, Cao BY, Guo ZY, 1–6. Copyright (2015), with permission from Elsevier

cross-plane conduction, the heat flux in the whole system is approximately q_B. From Eq. 6.68, one has

$$\lim_{l \to 0} \kappa_{y,\mathrm{eff}} = \frac{l}{2} \frac{\kappa_0}{\lambda_{\mathrm{R},0}}. \tag{6.72}$$

For the in-plane heat conduction, Eq. 6.48 yields the thin film limit as

$$\frac{l}{\lambda_{R,0}} \kappa_0 \nabla T = -\boldsymbol{q}(r) + \frac{l^2}{5} \nabla^2 \boldsymbol{q}(r). \tag{6.73}$$

The solution of Eq. 6.73 is

$$\lim_{l \to 0} \kappa_{x,\mathrm{eff}} = \frac{\kappa_0 l}{\lambda_{\mathrm{R},0}} \left[1 - \frac{2}{\sqrt{5}} \tan h \left(\frac{\sqrt{5}}{2} \right) \right]. \tag{6.74}$$

From Eqs. 6.72 and 6.74, one can obtain

$$\lim_{l \to 0} A^{\mathrm{nf}} = \frac{1}{2\left[1 - \frac{2}{\sqrt{5}} \tan h \left(\frac{\sqrt{5}}{2} \right) \right]} \approx 1.8. \tag{6.75}$$

It is also shown by Eqs. 6.72 and 6.74 that the in-plane and cross-plane thermal conductivities are both linear to the film thickness in the thin film limit, which agrees with the experiment and numerical observations.

The heterogeneous thermal conductivity also exists in traditional materials. However, such heterogeneity is normally caused by the heterogeneous crystallinity or nonuniform material composite. In contrast, the heterogeneity shown in Fig. 6.12 is induced by the ballistic phonon transport in nanofilms. This mechanism is different from traditional heterogeneities, which can be called the size effect-induced heterogeneous thermal conductivity.

6.5 Conclusion

In bulk material, the derivation of the general heat conduction law based on the thermomass theory assumes a linear dependence of the friction force on the drift velocity. In nanosystems, the boundary scattering of phonons plays an important role, causing a different behavior from bulk materials. Therefore, the general heat conduction law should be further modified/extended to describe the size effect of the effective thermal conductivity of nanosystems. The phonon gas viscosity is defined in analogy to the Darcy–Brinkman flow in porous medium. The phonon gas viscosity is proportional to the MFPs of *N* processes. In the balance equation of phonon gas momentum, the friction comes from both the flow region and the boundary. The viscosity allows the boundary to impede the flow additionally. The heat flux is nonuniform at the cross section due to the viscosity. The near boundary region has smaller heat fluxes. The boundary effect attenuates with a characteristic length, l_B. Therefore, the nonnegligible boundary friction in small-sized materials reduces the effective thermal conductivity.

In nanosystems, the system size is comparable with phonon MFPs. The boundary scattering reduces the effective phonon MFPs. The reduction of MFPs of *N* processes induces the viscosity decrease. This behavior is like the rarefication effect happening in micro-channel gas flow. The reduction of MFPs of R processes affects the permeability of phonon gas flow. The modification algorithm is proposed to assess the effective MFPs, which allows considering the viscosity and rarefication effects simultaneously. Thereby the prediction models for the in-plane thermal conductivity of nanofilms and nanowires are established. The size-dependent effective thermal conductivity predicted by the present model agrees well with experiments for both nanofilms and nanowires.

In the cross-plane heat conduction, the near boundary phonon distribution function is mainly determined by the wall properties. It is a ballistic distribution, f_B. f_B gradually relaxes to the diffusive distribution, f_B, due to scattering. Therefore, the boundary condition for the one-dimensional heat conduction is proposed, i.e., $f = f_B$ at the boundary. Inserting this condition into the phonon Boltzmann equation gives the ballistic-diffusive model for the cross-plane heat conduction. The material properties and MFP modification algorithm are selected the same as the in-plane problem. The prediction for the room temperature Si nanofilm agrees well with the

MD results. The in-plane thermal conductivity is different from the cross-plane thermal conductivity. The difference is induced by the size effect of nanosystems. When the system size grows to the bulk limit, the thermal conductivity of film tends to be homogeneous.

References

1. Majumdar A (1993) Microscale heat-conduction in dielectric thin-films. J Heat Transf 115 (1):7–16
2. McGaughey AJH, Landry ES, Sellan DP et al (2011) Size-dependent model for thin film and nanowire thermal conductivity. Appl Phys Lett 99(13):131904
3. Chen G (1998) Thermal conductivity and ballistic-phonon transport in the cross-plane direction of superlattices. Phys Rev B 57(23):14958
4. Yang R, Chen G (2004) Thermal conductivity modeling of periodic two-dimensional nanocomposites. Phys Rev B 69(19):195316
5. Dames C, Chen G (2003) Theoretical phonon thermal conductivity of Si/Ge superlattice nanowires. J Appl Phys 95(2):682–693
6. Alvarez FX, Jou D (2007) Memory and nonlocal effects in heat transport: from diffusive to ballistic regimes. Appl Phys Lett 90(8):083109
7. Alvarez FX, Jou D (2008) Size and frequency dependence of effective thermal conductivity in nanosystems. J Appl Phys 103(9):094321
8. Guyer RA, Krumhansl JA (1966) Solution of the linearized phonon Boltzmann equation. Phys Rev 148(2):766–778
9. Guyer RA, Krumhansl JA (1966) Thermal conductivity, second sound, and phonon hydrodynamic phenomena in nonmetallic crystals. Phys Rev 148(2):778
10. Sellitto A, Alvarez FX, Jou D (2010) Second law of thermodynamics and phonon-boundary conditions in nanowires. J Appl Phys 107(6):064302
11. Jou D, Criado-Sancho M, Casas-Vázquez J (2010) Heat fluctuations and phonon hydrodynamics in nanowires. J Appl Phys 107(8):084302
12. Jou D, Cimmelli VA, Sellitto A (2012) Nonlocal heat transport with phonons and electrons: application to metallic nanowires. Int J Heat Mass Transf 55(9):2338–2344
13. Sellitto A, Alvarez FX, Jou D (2012) Geometrical dependence of thermal conductivity in elliptical and rectangular nanowires. Int J Heat Mass Transf 55(11):3114–3120
14. Ma Y (2012) Size-dependent thermal conductivity in nanosystems based on non-Fourier heat transfer. Appl Phys Lett 101(21):211905
15. Beavers GS, Joseph DD (1967) Boundary conditions at a naturally permeable wall. J Fluid Mech 30(01):197–207
16. Brinkman HC (1949) A calculation of the viscous force exerted by a flowing fluid on a dense swarm of particles. Appl Sci Res 1(1):27–34
17. Nield DA, Bejan A (2006) Convection in porous media, 3rd edn. Springer, New York
18. Stops DW (1970) The mean free path of gas molecules in the transition regime. J Phys D Appl Phys 3(5):685
19. Asheghi M, Leung YK, Wong SS et al (1997) Phonon-boundary scattering in thin silicon layers. Appl Phys Lett 71(13):1798–1800
20. Ju YS, Goodson KE (1999) Phonon scattering in silicon films with thickness of order 100 nm. Appl Phys Lett 74(20):3005–3007
21. Liu W, Asheghi M (2004) Phonon–boundary scattering in ultrathin single-crystal silicon layers. Appl Phys Lett 84(19):3819–3821
22. Ju YS (2005) Phonon heat transport in silicon nanostructures. Appl Phys Lett 87(15):153106

23. Li D, Wu Y, Kim P et al (2003) Thermal conductivity of individual silicon nanowires. Appl Phys Lett 83(14):2934–2936
24. Moore AL, Saha SK, Prasher RS et al (2008) Phonon backscattering and thermal conductivity suppression in sawtooth nanowires. Appl Phys Lett 93(8):083112
25. He Y, Galli G (2012) Microscopic origin of the reduced thermal conductivity of silicon nanowires. Phys Rev Lett 108(21):215901
26. Esfarjani K, Chen G, Stokes HT (2011) Heat transport in silicon from first-principles calculations. Phys Rev B 84(8):085204
27. Mingo N (2003) Calculation of Si nanowire thermal conductivity using complete phonon dispersion relations. Phys Rev B 68(11):113308
28. Veijola T, Turowski M (2001) Compact damping models for laterally moving microstructures with gas-rarefaction effects. J Microelectromech Syst 10(2):263–273
29. Guo ZL, Shi BC, Zheng CG (2007) An extended Navier-Stokes formulation for gas flows in the Knudsen layer near a wall. Europhys Lett 80(2):24001
30. Feng XL (2001) Molecular dynamics study on lattice thermal conductivity of nanoscale thin films. Doctoral dissertation, Tsinghua University, Beijing
31. Wang ZH, Li ZX (2006) Research on the out-of-plane thermal conductivity of nanometer silicon film. Thin Solid Films 515(4):2203–2206
32. Chen G (2001) Ballistic-diffusive heat-conduction equations. Phys Rev Lett 86(11):2297–2300

Chapter 7
Conclusion

The salient content of this work can be concluded as the following:

1. The microscopic foundation of the general heat conduction law based on thermomass theory is investigated. Based on the thermomass theory, the phonon gas in dielectric materials has its equivalent mass. The momentum balance equation of phonon gas gives the general heat conduction law. Based on the phonon Boltzmann equation, the momentum balance equation of phonon gas based on the thermomass theory is derived microscopically. The transient inertia of phonon gas arises from the first order expansion of the phonon state distribution function, while the spatial inertia comes from the second order expansion. If only the zeroth order of the distribution function is considered, the inertia effect of phonon gas motion is thereby neglected. In this case the general heat conduction law reduces to the traditional Fourier's law.
2. The general expression of entropy production is established based on the dynamical analysis of non-Fourier heat conduction. The entropy production in the classical irreversible thermodynamics is not applicable for the non-Fourier heat conduction. Based on the thermomass theory, the generalized force in heat conduction should be the friction force rather than the driving force. The generalized flux is proportional to the drift velocity of thermomass. Therefore, the entropy production is proportional to the dissipation of thermomass energy, which is the product of the friction force and the drift velocity. The modified general entropy production is applicable for both the Fourier heat conduction and the non-Fourier heat conduction.
3. The nonequilibrium and equilibrium temperatures in the non-Fourier heat conduction are actually the static and stagnant pressures of the phonon gas in dielectric solids. The phonon gas motion in heat conduction has its kinetic energy, which makes a difference between its static pressure and stagnant pressure. The static pressure of phonon gas corresponds to the static temperature. It is the real thermodynamic state variable. The stagnant pressure of phonon gas represents the sum of the potential and kinetic energies of phonon gas. In the irreversible transport, the temperature cannot be defined based on the total energy of particles. Instead, the contribution from the kinetic energy of phonon gas should be excluded. The static temperature of phonon gas is the

Y. Dong, *Dynamical Analysis of Non-Fourier Heat Conduction and Its Application in Nanosystems*, Springer Theses,
DOI 10.1007/978-3-662-48485-2_7

criterion of thermodynamic equilibrium. Defined based on the static temperature, the relation between the internal energy, entropy, and temperature is identical to the traditional form.

4. The Onsager reciprocal relation is derived based on the dynamical analysis. The traditional proof of Onsager reciprocal relation requires that the generalized flux can be expressed by the time derivative of state variables. However, the state variables corresponding to the heat flux are ill defined. Based on the thermomass theory, the linear regression of fluctuation assumed by Onsager is the balance of thermomass inertia and friction forces. The time derivative of heat flux corresponds to the inertia of thermomass. Therefore, the average displacement in the microscopic fluctuation of thermomass is the eligible state variable satisfying Onsager's requirement. From a dynamical viewpoint, the Onsager reciprocal relation is essentially the symmetry of the parameter matrix among interaction forces and transport fluxes. A macroscopic derivation of Onsager reciprocal relation is thereby developed based on the Galilean invariance and Newton's third law of dynamics (momentum conservation principle).
5. The general heat conduction law is further extended to describe the size effect of the thermal conductivity of nanosystems. The boundary resistance is significant in nanosystems. The viscosity of phonon gas is proposed in analogy to the porous flow. The phonon gas flow in nanosystems is impeded not only by the scattering inside the medium but also by boundary scattering, which is modeled by a viscous effect of the phonon gas. The boundary resistance becomes larger in smaller systems, leading to the decrease of effective thermal conductivity in nanosystems. The viscous and rarefication effects of phonon gas flow in nanosystems are analyzed to establish the prediction models for the in-plane heat conduction in nanofilms and nanowires. The prediction agrees well with experiments. For cross-plane heat conduction, the phonons in the near boundary region are highly nonequilibrium. The ballistic-diffusive model for the cross-plane heat conduction in nanofilms is established based on the phonon Boltzmann equation. This model yields good agreement with the MD results. The in-plane and cross-plane thermal conductivities are different due to the heterogeneous phonon-boundary scattering. It is predicted that the ratio of the in-plane thermal conductivity over the cross-plane thermal conductivity is 1.8 in the thin film limit. Such heterogeneity gradually vanishes with increase of film thickness.

MIX
Papier aus verantwortungsvollen Quellen
Paper from responsible sources
FSC® C105338

FSC
www.fsc.org

If you have any concerns about our products,
you can contact us on
ProductSafety@springernature.com

In case Publisher is established outside the EU,
the EU authorized representative is:
Springer Nature Customer Service Center GmbH
Europaplatz 3, 69115 Heidelberg, Germany

Printed by Libri Plureos GmbH
in Hamburg, Germany